# L KINGDOM

This classification shows the major groups within the classes – mammals, reptiles, amphibians, birds, fish and insects – that make up the animal kingdom.



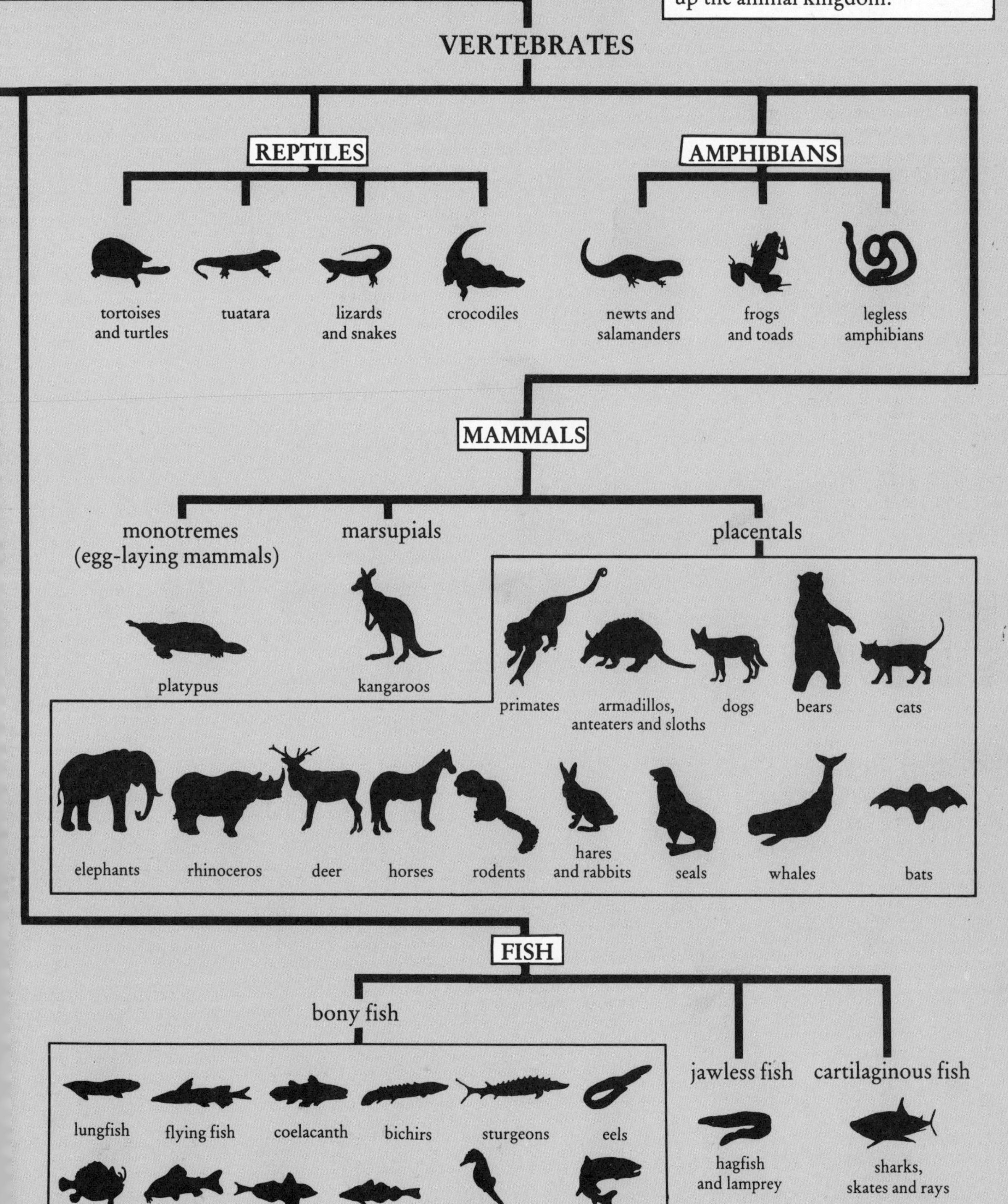

**Written by**
Stephen Attmore

**Illustrated by**
Graham Allen
Bob Bampton
Ray Cresswell
John Francis
Bob Hersey
Alan Male
Colin Newman
John Rignall
David Thompson
Phil Weare

**Front cover illustration by**
Jim Channell

**Edited by**
Trevor Weston

ISBN 0 86112 425 1

Published by Brimax Books, Newmarket, England 1987
Printed in Hong Kong

# ANIMAL ENCYCLOPEDIA

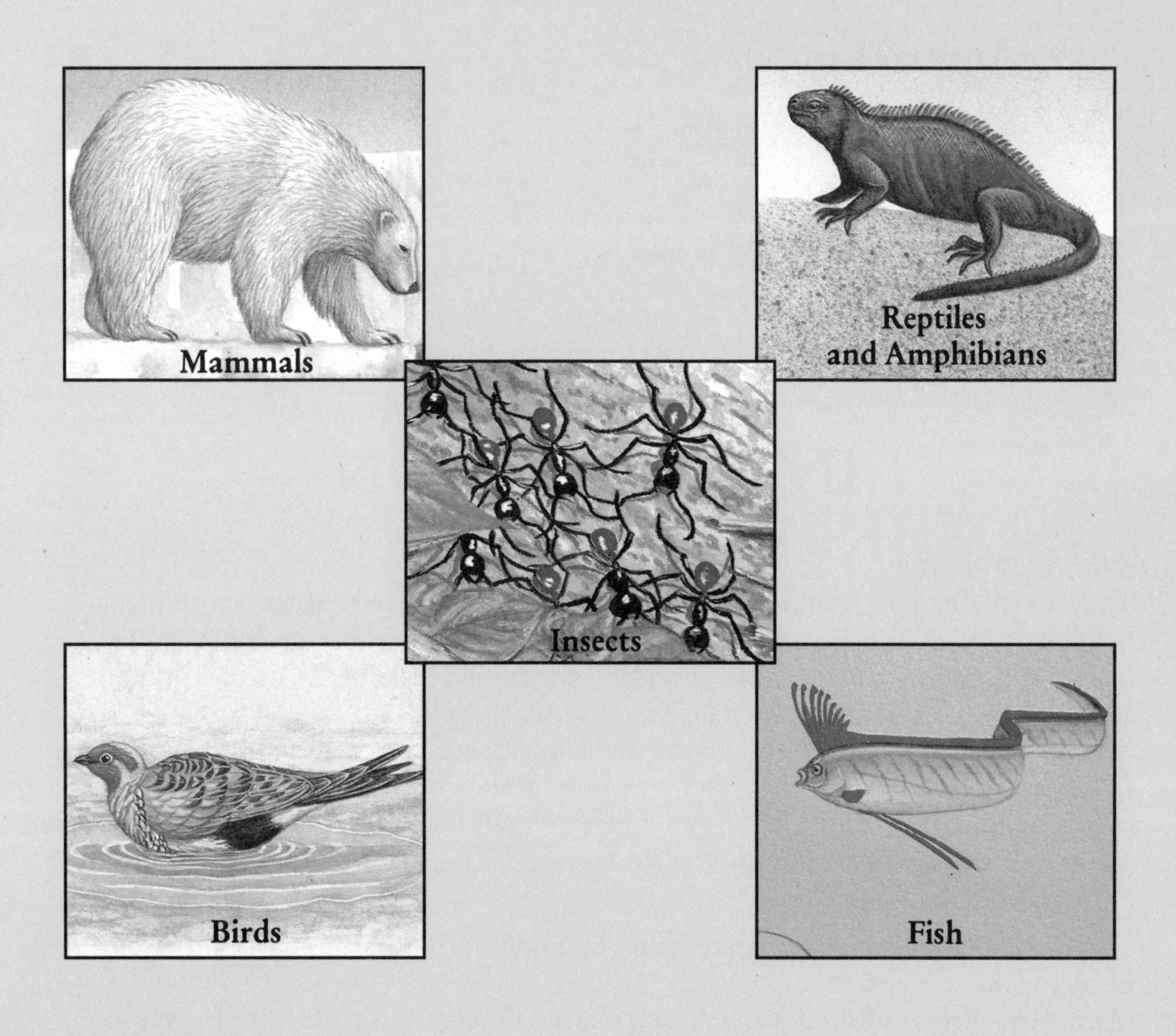


BRIMAX BOOKS · NEWMARKET · ENGLAND

# CONTENTS



# INTRODUCTION

*Animal grouping*

The animal kingdom is split into six main classes: mammals; reptiles; amphibians; fish; birds and invertebrates. Invertebrates are creatures without backbones, like insects and spiders. This is the largest class, containing 95% of the species in the animal world. The *Animal Encyclopedia* is similarly split into classes, though for ease of reference, the invertebrates are divided between land invertebrates in the *Insects* section and sea invertebrates which are in the *Fish* section. There is also a fascinating final section which shows where animals live (desert, jungle, forest, sea etc.) and how they depend on one another to survive. This section is called *Animal Links*.

*Colour tabs*

There is a colour-coding system to help you find your way around the book.
Look for the corner tabs:
Mammals – Brown; Reptiles and Amphibians – Orange; Fish – Green; Insects – Blue; Birds – Yellow; Animal Links – Red.
There are also colour panels within each section containing further, general information about particular animal groups and species.

*Sizes and scale*

Each animal in the Encyclopedia has its own entry and illustration. The individual entries contain basic information such as the animal's size, what it eats and how many babies it has, as well as many fascinating facts. The sizes are given as approximate total lengths unless stated otherwise: in birds that is the length from bill tip to tail tip; in tortoises and turtles length of shell is given; in dogs the height given is to the shoulder. The vast range of sizes means that the illustrations cannot all be drawn to a single scale but every effort has been made to ensure that animals on a particular page are drawn in rough proportion to one another.

*World map*

The entries state where each animal is to be found. A map is provided on pages 152–153 so that you can identify where a particular animal lives. The map also gives some examples of animal species that are in danger of becoming extinct.

*Terminology*

In the Encyclopedia the terms 'species', 'family', 'order' and 'class' are used. To explain the difference, let's take an example: the red kangaroo is a 'species' of the kangaroo 'family'. Kangaroos, as well as wombats, koalas etc., belong to the 'order' marsupials. Marsupials are in the 'class' called mammals. An animal 'species' is defined by those animals in a group that can breed with one another.

*Glossary*

Important technical words, such as prehensile, arthropod and so on, used in the entries, are explained in a glossary at the back of the book. So, if you are unsure of the meaning of a word, look in the glossary.

*Use of bold and italic*

Where an animal's name appears in **bold** letters within the text, this indicates the particular species of animal that has been illustrated. For example, it is the **pygmy marmoset** that is illustrated under the marmoset entry. Where an animal's name appears in *italics* within the text, this means the animal has its own entry elsewhere in the Encyclopedia. You can find it by looking in the index.

*Classification chart*

The classification chart in the front and back of the book shows the structure of the animal kingdom. There are over forty-three thousand species of vertebrate and over one million species of invertebrate in all. This Encyclopedia cannot include every one but it gives examples from all the major groups. You will find common animal families, like cats and dogs, as well as rare species such as the takahe and the giant panda. You will discover some fascinating information about familiar animals; you may also find out about animals that you never knew existed.

# MAMMALS

There are over 4000 species of mammal on our planet today. They have all evolved from reptiles and they have adapted to most environments, including the sea, the trees and even the air in the form of bats. They can be divided into three categories: monotremes (the platypus and spiny anteaters) have retained the egg-laying habits of their ancestors, the reptiles; marsupials (e.g. kangaroos and wallabies) whose developing young break out of their shell and crawl into the mother's pouch to suckle; and placentals (including rodents, humans and whales) whose offspring remain inside the female and are born at an advanced stage of development.

A mammal is a vertebrate (an animal with a backbone); it is warm-blooded and usually hairy. Its body temperature is controlled by a brain mechanism and is kept constant by body hair and the evaporation of sweat. Some mammals hibernate during long spells of cold weather. The animal goes into a deep sleep and its body temperature drops down close to that of its surroundings. Its heart beat and breathing slow dramatically so that it uses the minimum of energy. Whilst hibernating, a mammal can survive for several months on stored body fat.

Nearly half the living species of mammal are rodents (e.g. rats and beavers) and almost a quarter are bats.

## Primates

These form the order of mammals to which you and I belong. Monkeys, apes and lemurs are also primates. Apart from human beings, who live worldwide, most primates are found in tropical regions such as Central and South America, Africa and southern Asia. They are tree-dwellers and have five digits on each of four limbs with nails instead of claws. They have a more rounded skull and a relatively larger brain than other mammals. The main difference between monkeys and apes is that monkeys have tails.

1 **Orang-utan** This giant ape with sparse, shaggy red hair gets its name from the Malay word for 'man of the woods'. A big male orang stands 1.5 m (5 feet) tall, has an arm span of 2.5 m (8 feet) and weighs up to 200 kg (440 lb). A female is only half this size and weight. Adults have a large, fatty throat pouch. These 'four-handed' apes use one hand and both feet to cling to branches whilst feeding on fruit, leaves, seeds and young birds. They climb down one tree and up another because if they jump, branches often break. At night an orang-utan quickly builds a simple nest of sticks. It usually lives alone but calls to others by making a smacking sound, like a kiss.

2 **Colobus monkeys** of Africa have long tails and long hind limbs which help them to leap from tree to tree. Even when they cannot be seen from the ground, these robust monkeys can be heard crashing through the tree-tops. They have only four fingers on each hand. The **Angolan black-and-white colobus** feeds mainly on leaves, fruit and insects. It is identified by the very long white hairs on its shoulders. The adult male grows to a length of 60 cm (2 feet) and has a 90 cm (3 foot) tail.

3 **Gibbons** are the smallest of the ape family. The smaller gibbons of Indonesia and south-east Asia are the fastest moving primates in trees. They reach speeds of over 32 km/h (20 mph) as they swing by their long arms from branch to branch. When they walk upright on the ground, gibbons hold their arms up high for balance. **Lar gibbons** feed in the upper branches of trees on fruit, leaves, shoots, flowers and occasionally insects. The female gives birth to a single, hairless baby which she keeps warm between her thighs and belly. Adult males grow to nearly 60 cm (2 feet).

4 **Diana monkeys** are members of the guenon family. They have very striking markings – a black and white face, white beard and chest, and distinctive chestnut patches on the back and hind limbs. There are also stripes on each thigh. Females are smaller than males, but look similar. They are found mostly in rain forests. They are very skilful climbers and live in troops of up to 30. They are most active in the early morning and late afternoon. These noisy and inquisitive animals feed on plants, insects and sometimes birds' eggs. They are about 60 cm (2 feet) long with an even longer tail.

5 **Red-bellied monkeys,** from Nigeria in Africa, are members of the guenon family. These rare monkeys have a dark face fringed with white side-whiskers. Their bellies and chests are usually red. Some, though, have grey bellies. Few specimens of this monkey have been found so little is known of their habits.

◀ **Mandrill** This African forest *baboon* is the largest and strongest member of the monkey tribe. Male mandrills have scarlet and blue ridges on their faces which make them look frightening to other animals. They weigh up to 54 kg (119 lb) and are 1 m (3¼ feet) long with only a stump for a tail. Females are duller and much smaller. Mandrills live in groups of 20–50 and walk on their fingers and toes. They are mainly vegetarian but supplement their diet of fruit and leaves with small animals which they find by overturning stones and dead wood. A single baby is born at any time of the year. Young mandrills are sometimes caught and eaten by *leopards.*

▶ **Douroucouli** This nocturnal primate is also known as the night or owl monkey because of its large eyes which enable it to see very well in the dark. Douroucoulis are related to the *capuchin, saki* and *uakari.* They rarely come down to the ground and pairs sleep in hollow trees or nest among foliage. They emerge at dusk to feed on fruit, leaves, insects, spiders and some small mammals. Males defend territory by 'boxing' with their rivals. An inflatable sac under the chin makes their call louder. Adult douroucoulis are up to 38 cm (15 inches) long, with a slightly longer tail. Their bushy fur is greyish with black and white face markings.

◀ **Capuchins** are lively, intelligent monkeys found in forests in South America and Trinidad. The **white-throated capuchin** is also known as the ring-tail because the end of its tail is coiled. It moves quickly from branch to branch and sometimes climbs down to the ground and crosses open country. Capuchins are inquisitive and investigate all plants and fruit in the hope that they are edible. They use stones to crack open nutshells. Adult capuchins grow to 38 cm (15 inches) and their tails are even longer. They often live for over 45 years. A baby capuchin clings with its hands and feet to its mother's belly.

▶ **Gorillas** are the largest and heaviest primates. The average height for an adult male is 1.7 m (5½ feet). Gorillas in the wild weigh about 220 kg (484 lb) but in captivity they weigh as much as 310 kg (682 lb). Troops of 16–30, led by a dominant male, spend most of the time on the ground walking on all fours. They feed only on vegetation. Gorillas do not need to drink water as there is enough juice in their food. Each gorilla builds a new nest every night. Young gorillas sleep with their mother until they are three. Gorillas are only dangerous when wounded or alarmed. When faced with an intruder, a dominant male beats his chest with both hands. He rushes through the undergrowth, breaking branches and thumping the ground with his hands. The gorilla's only enemy is the upright primate, the human.

▼ **Uakari** (pronounced wakari) Related to the *saki*, this South American monkey has long, coarse hair and a beard. The face and head of the **bald uakari** are nearly hairless. Small troops are active during the day, feeding on fruit, leaves, insects, small mammals and birds. These agile climbers are rarely seen in the wild. They live in the tree-tops and seldom descend to the ground. The three species of uakari have tails shorter than their bodies. The bald uakari is no more than 57 cm (23 inches) long, with a 15 cm (6 inch) tail.

▼ **Tamarins** and *marmosets* are known as flat-nosed monkeys. They do not swing from branch to branch like most primates, but jump swiftly through the trees. **Golden lion tamarins** are an endangered species. It is a beautiful animal with a silky, golden mane. Its diet consists of fruit, insects, lizards, small birds and eggs. Golden lion tamarins are noisy monkeys, threatening outsiders with a high-pitched twittering. Females give birth to one or two young. The father helps to look after the babies.

▼ **Siamangs**, largest of the gibbons, live in mountain forests in Malaysia and Sumatra. They swing from branch to branch in search of fruit, shoots, insects and birds' eggs. Sometimes they walk upright on strong branches. Their fur is entirely black and they have webbed skin between the second and third toes of each foot. Newborn siamangs are almost hairless. Families call to one another with alternate booms and shrieks made louder by a large throat sac.

▼ **Macaques** are monkeys related to *baboons* and *mandrills*. They are among the most intelligent of all primates. Nearly all 60 species of macaque spend a lot of time on the ground. They communicate by making a variety of visual and vocal signals. The **Japanese macaque**, the only wild monkey in Japan, is the only primate (apart from humans) able to survive in near-freezing temperatures. Their shaggy coat helps them to keep warm. They like to bathe in hot springs.

▼ **Marmosets** are the smallest of all monkeys. They live and move like *squirrels* – using their long, sharp claws as they move in short dashes along branches. Marmosets are different from other monkeys because they have claws and wisdom teeth. Their diet is mainly nuts, insects and tree sap. The **pygmy marmoset** from South America is only 15 cm (6 inches) long with a 17 cm (7 inch) tail. Marmosets spend most of their time in trees. At night they sleep curled up in holes. They are sometimes attacked by large birds of prey.

▼ **Baboons** are species of monkey found mainly in Africa. **Olive baboons** live in large troops of up to 150. Each troop sleeps in its own home range of trees. They travel in a long procession to feeding grounds in the morning. They eat fruit, grass, insects and lizards. Some troops kill hares and young antelope. A young olive baboon clings to its mother's belly for the first five weeks. Then it rides on her back until it is six months old. Adult olive baboons grow to 1 m (3¼ feet) and have tails as long as 75 cm (2½ feet).

▼ **Proboscis monkeys** live in Borneo. They get their name from the male's long, bulbous nose. When the male proboscis makes a honking call, his nose straightens out. He is much larger than the female and about twice her weight. Proboscis monkeys run and leap from branch to branch of mangrove trees, using their 75 cm (2½ foot) tail to help them balance. Their long fingers and toes help them to grip. They are most active in the morning, feeding on mangrove leaves and shoots, pedada trees and other flowers. The rest of the day is spent sunbathing in the tree-tops. Proboscis monkeys are becoming rare in the wild and they are difficult to keep in captivity.

▼ **Sakis** are South American monkeys related to *uakaris, howlers, capuchins* and woolly monkeys. The three species of saki have long, coarse hair framing their faces and bushy tails up to 50 cm (20 inches) long. They grasp objects between their second and third fingers. The **monk saki** is a shy animal which lives high in trees and never comes down to the ground. It moves on all fours but sometimes stands upright before leaping to another tree. Monk sakis live in pairs or small family groups. They eat fruit, berries, honey, leaves, bats, mice and birds. They drink by dipping the fur on the back of their hands in water and licking it. The black-bearded saki is an endangered species because the trees where it lives are being chopped down.

▼ **Lemurs** are agile tree-dwellers found in Madagascar, off the coast of South Africa. They are related to monkeys, apes and human beings. It is possible that lemurs once roamed Africa, Europe and North America until monkeys evolved and drove them out. Lemurs eat fruit and leaves from the tamarind tree. Some species also eat insects and other small animals. The **ring-tailed lemur** is about the size of a cat – 45 cm (1½ feet) – with a long tail – 55 cm (1¾ feet). It has a pointed muzzle, large eyes and triangular ears. The female carries her young in her mouth. All lemurs scrape and groom their fur using lower front teeth shaped like a comb. The second toe on each foot has a claw which they use as an earpick. Male lemurs have stink fights. They rub their scent on twigs and wave them about.

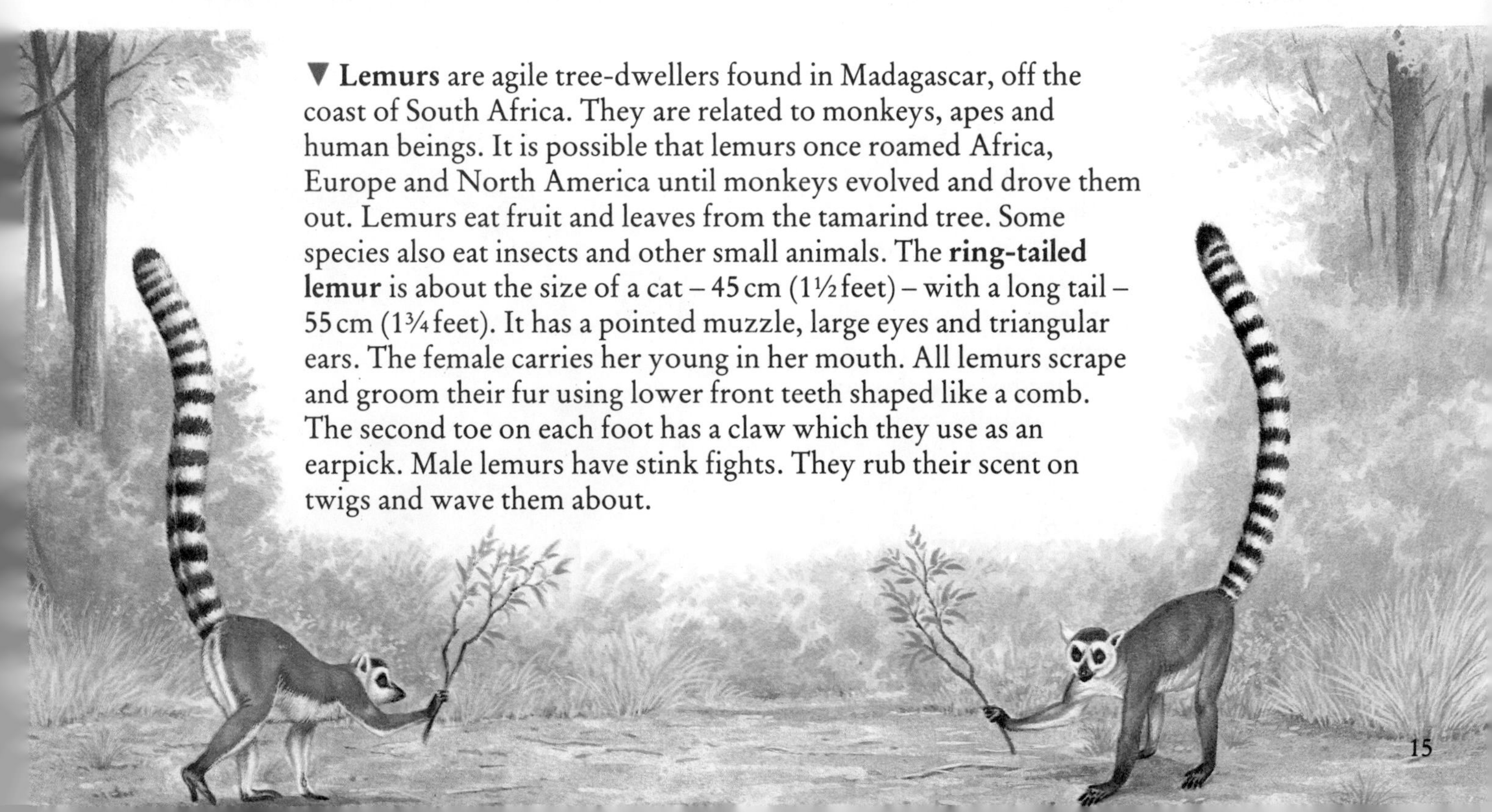

► **Howler monkey** The name of this South American relative of the *spider monkey* is very apt because its cry can be heard up to 3 km (1¾ miles) away. It is said to be the loudest noise produced by any animal. Howlers howl to attract others to food and when competing troops meet. A special bone makes the neck swell and forms a large voice-box. Although they draw attention to themselves with their howls, no predator can reach the howler monkeys in the tree-tops. The **red howler** is 90 cm (3 feet) long, with a tail of equal length which has a sensitive, naked area on the underside.

◄ **Loris** This primate family lives in Africa, India and south-east Asia. They are related to *bushbabies.* The loris is a nocturnal tree-dweller with long, thin limbs. It creeps up undetected on its prey and quickly grabs it with both hands. A loris can grip branches for long periods without tiring. A **slender loris** weighs about 200 g (7 oz) and is only 25 cm (10 inches) long. It has no tail and sleeps during the day rolled up in a ball in a tree. At night it feeds on insects (especially grasshoppers), lizards, small birds and vegetation. The slender loris is found in southern India and Sri Lanka.

► **Spider monkeys** are found in the topmost branches of South American forests. They have very long legs and walk upright on the ground, on all fours along branches and they swing through the trees. Groups of up to 100 sleep together at night, but separate when feeding. Spider monkeys communicate with sharp barks. They growl at intruders and shower them with branches. The **woolly spider monkey** is 60 cm (2 feet) long, with a tail of 75 cm (2½ feet). Its facial skin goes red when it is excited. When on the move, a female spider monkey bridges the gap between two trees by gripping one branch with her hands and the other with her feet and tail. Her young then scramble across her body.

◀ **Aye-aye** This rare primate was at one time found in forests in Madagascar, but is now only seen in nature reserves. A relative of the *lemur*, this nocturnal creature is well adapted for tree life. It has long, slender fingers with a very long middle one. The aye-aye uses its long fingers to tap on bark to locate insects. Then it uses its huge, bat-like ears to listen for the grub's movements. It gnaws a hole through the bark before winkling out its prey with its middle finger. A female gives birth to a single baby every two or three years. Some males grow to a length of 45 cm (1½ feet) and have a 60 cm (2 foot) tail.

▶ **Tarsiers** are the only primates whose diet is entirely carnivorous. They eat insects, lizards, spiders and mice at night. Found in south-east Asia, the tarsier is well adapted for clinging and jumping: its long hind legs help it to leap gaps of 1.8 m (6 feet); its tail acts as a brake and its large eyes enable it to judge distances; suckers on the tips of fingers and toes give a firm grip when clinging to stems. The tarsier keeps one eye open when resting. It can turn its head through 180 degrees to watch for danger and it twists its thin ears to focus on sound. The **western tarsier** is no more than 15 cm (6 inches) long and can weigh as little as 85 g (3 oz).

◀ **Chimpanzee** The most intelligent primate (apart from humans), this African ape uses a wide variety of calls, gestures and facial expressions to communicate. It is a tool-user, wielding stones as weapons and using sticks for probing ants' nests. Chimps prefer to travel on the ground, walking on all fours. They climb trees to pick and eat ripe fruit and leaves. Some eat ants, small monkeys and pigs. When chimps want to drink they dip a hand in water and lick the water off. At dusk they build a new tree nest and sleep for 12 hours. Males of equal rank spend hours each day grooming one another. Adults are about 1 m (3¼ feet) tall and can live for 50 years.

◀ **Bushbabies** are named after the noise they make. They sound like a young child crying. They are nocturnal mammals and live in Africa, high up in trees. They use their powerful hind legs to leap from branch to branch. Bushbabies have opposable thumbs and toes which enable them to hold on to branches. The **greater bushbaby** suddenly pounces on its prey, killing it with a bite. It eats insects, reptiles, birds and plant material. The greater bushbaby's tail – up to 52 cm (21 inches) long – is longer than its body. The smallest species is Demidoff's bushbaby which weighs only 70 g (2½ oz).

▶ **Indri** are an endangered species related to *lemurs*. They have a stumpy tail and long legs – from head to toe they measure nearly 1 m (3¼ feet). Indris are magnificent jumpers. They use their powerful hind legs to push off and they travel through the air with body upright. They bounce from tree to tree and come down to the ground very rarely. Morning and evening, a wailing cry can be heard high in the forests of Madagascar. An indri can continue wailing for several minutes without taking in air. All this noise draws attention, but they are so high up that no large predator can get near them. Indris are boldly marked in black and white. They have a reflecting layer in their eyes. This is common in nocturnal animals so it is probable that indris were once night animals. They live in family groups and eat leaves, shoots and fruit.

◀ **Tree shrews** have lived on Earth for at least 70 million years. There are 15 species of tree shrew living in the Asian forests. The exception is the Philippine tree shrew which lives on the ground. The female **common tree shrew** builds a nest in a fallen tree or among roots. She feeds her young once every two days. They are active in the daytime, feeding on insects and seeds. Sharp, curved claws are used to dig out insects. Their forepaws have knobbly pads (see diagram). Tree shrews drink frequently and are fond of bathing. The pen-tailed tree shrew is the smallest known primate. This rare species weighs under 50 g (1¾ oz) and its body is only 10 cm (4 inches) long.

▶ **Cuscus** This tree-living marsupial is a relative of the *kangaroo.* It is found in Australia and New Zealand. A cuscus chatters loudly and gives off a horrible smell if it is attacked. It is rarely seen during the day, keeping still among thick foliage. At night it feeds on leaves, insects and sometimes birds' eggs. When feeding, it grips the branch with its back feet and tail so that its front feet are free for holding food. The **spotted cuscus** is about 90 cm (3 feet) long – the size of a domestic cat.

◀ **Opossums** are the only marsupials (mammals who carry their young in a pouch) outside Australia. They are found in North and South America. The smallest mammal at birth is the mouse opossum – no longer than a grain of rice. **Virginia opossums** are the largest of the opossum family – adult males are up to 50 cm (20 inches) long with a thin, naked tail the same length. They will eat anything they can find in the undergrowth. They are slow on the ground but are expert climbers. When disturbed, they will 'play dead' and give off a nasty smell. Of a litter of 8–18 only about 7 will survive pouch life.

▶ **Skunk** This mammal has a highly successful way of defending itself: it raises its tail and sprays two jets of stinking fluid at its enemy which causes temporary blindness. Related to *badgers, weasels* and *otters,* this nocturnal carnivore eats rodents, insects, birds' eggs and vegetation. The **western spotted skunk** is found in Central and South America. No two skunks have exactly the same markings. They live in dens underground, but are good climbers and sometimes shelter in trees. An adult is about 55 cm (22 inches) long, including the bushy tail. The young skunk, called a kit, is born blind, hairless and toothless and weighs 28 g (1 oz). Its eyes open at 21 days.

◀ **Kangaroos** Over 50 species of this vegetarian marsupial are found in Australia. The name means 'large-footed'. Kangaroos move in great bounds, propelled by their powerful hind legs. Their sturdy tail acts as a counterbalance when leaping and as a prop when resting. Over long distances they average a speed of 40 km/h (25 mph). Males grow throughout their lives and male **red kangaroos** weigh as much as 70 kg (154 lb). They stand over 1.5 m (5 feet) tall, with a 1 m (3¼ foot) tail. A newborn joey (baby kangaroo) weighs only 0.75 g (1/40 oz). Using its front claws, it clambers into its mother's pouch where it feeds and grows for 240 days.

▶ **Koalas** This Australian marsupial feeds only on the leaves of eucalyptus trees. An adult koala consumes over 1 kg (2¼ lb) of leaves in one night. During the day it sleeps curled up in a tree-fork. Koalas only climb down to the ground to move on to a fresh tree. They lick the ground, probably as an aid to digestion. A koala has a soft pad on its nose. It is 85 cm (2¾ feet) long and has no tail. A baby koala is less than 2.5 cm (1 inch) long when it crawls into its mother's rearward-opening pouch. It emerges after 5 to 6 months and clings to its mother's back. Many koalas suffer from a disease which affects their lungs, joints and brain and leaves them unable to have babies.

◀ **Duck-billed platypus** is an egg-laying mammal found in and near Australian and Tasmanian rivers. It has dense fur, webbed feet and a flat tail. Its 'bill' is a sensitive snout covered in soft skin and the nostrils at the end of its upper bill only function out of water. A platypus relies on its bill to find shrimps and insect larvae in the mud because its ears and eyes are closed by skin flaps when under water. This nocturnal feeder has a huge appetite. Females dig a 12 m (40 foot) burrow and lay two or three rubbery eggs.

▲ **Wombats** These nocturnal marsupials live in Australia and Tasmania. They dig wide, deep burrows at the base of trees or rocks. Some burrows are as long as 30 m (100 feet). When the mother wombat digs, the baby does not get covered in dirt because the pouch faces backwards. The young wombat stays in the pouch for several months. Wombats follow the same pathways each night, searching for edible grasses and roots. A male wombat grows to a length of 1.2 m (4 feet). The adult has only a trace of a tail left. Wombats live longer than other marsupials – often over 20 years. The oldest recorded age was 26 years 22 days.

▲ **Wallabies** These Australian marsupials are members of the *kangaroo* family. The **nail-tailed wallaby** get its name from a fingernail-like spur at the tip of its tail. They were thought to be extinct, but a few were discovered in 1974. Their decline was caused by competition from rabbits and sheep for scrubland food. Many were killed by foxes, dogs and human hunters. Male nail-tailed wallabies are up to 67 cm (27 inches) long with a tail of the same length. Female wallabies have one baby at a time. It is not known how long the baby stays in the mother's pouch.

◀ **Tasmanian devil** These marsupials lived in Australia until they were all killed by *dingos* They survive on Tasmania because there are no wild dogs there. Their name derives from their savagery towards other animals. They can smash through bones with their powerful jaws. They hunt at night, feeding on small wallabies, birds, fish and frogs. Their young (up to 4) are kept in a closed pouch for 15 weeks. In late September, feet or a tail may be seen sticking out of the pouch. Males grow to 75 cm (2½ feet) and have a 30 cm (1 foot) tail. They weigh about 9 kg (20 lb).

▶ **Honey possum** This tiny marsupial lives in Australia. It is the size of a mouse – about 8 cm (3 inches) long with a slightly longer tail. It has no close relative and is completely different from the *opossum*. Honey possums often hang upside down while feeding, holding on to a branch with their tail. Using their pointed jaws and long, bristly tongue they eat pollen and nectar from flowers. Baby honey possums remain in their mother's pouch until they are four months old.

► **Sloth** The slowest moving land mammal is the **three-toed sloth** whose usual speed is under 1.6 km/h (1 mph). This nocturnal herbivore is so well adapted to tree life in Central and South American rain forests that it is unable to walk on the ground. A sloth spends most of its life hanging upside down from branches by its curved claws. Females even give birth while hanging. Sloths' hair grows downward and each outer hair is grooved which enables rainwater to run off. Microscopic green algae grow in the grooves providing the sloth with camouflage. The three-toed sloth is 60 cm (2 feet) long. Two extra neck bones enable it to turn its head through 270°.

◄ **Pangolin** This scaly anteater is found in Africa and Asia. Its overlapping scales are movable and have sharp edges. They form a protective defence. When danger threatens, a female curls up cuddling her baby. African **tree pangolins** are adept climbers, using their long tails to grip branches. They eat only certain species of ants and termites. Tearing open nests with their powerful forelimbs, they sweep up the insects with a very long tongue. They have no teeth but grind food in a horny-surfaced stomach. Some tree pangolins grow to a length of 1 m (3¼ feet).

▼ **Anteater** All members of the American anteater family have extremely long snouts and no teeth. The tongue of a **giant anteater** is covered with saliva to which insects stick easily and it can extend it 60 cm (2 feet). The anteater breaks into an ants' nest using its powerful, clawed forefeet. The giant anteater is the largest of the family – an adult is often over 2 m (6½ feet) in length, from nose tip to the end of its long-haired tail. Unlike other anteaters, it does not climb trees.

◀ **Porcupine** This rodent with sharp-tipped spines moves with shuffling steps. North American porcupines have over 20,000 quills. The **crested porcupine**, found mainly in Africa, has quills up to 30 cm (1 foot) long. These hollow spines can be rattled as a warning. If the threat continues, it charges backwards and drives its spines into the enemy. It digs a burrow in which it spends the day, emerging at night to feed on green plants, roots and field crops. The crested porcupine rarely climbs trees. Males and females look alike and grow to 85 cm (2¾ feet). They produce two litters a year of two to four young, born with soft spines.

▶ **Hedgehog** This mammal gets its name from its pig-like habit of rooting around for food in hedgerows. When disturbed, it rolls itself into a tight ball, protecting its legs, head and belly. Its enemy is faced with a bundle of sharp spines. The **western European hedgehog** feeds on worms, insects, frogs and berries. It grunts as it shuffles along on its short legs. About 28 cm (11 inches) long, it weighs 1.2 kg (2½ lb). Each year many die trying to cross roads or are poisoned by chemicals used in pest control. In the northern part of its range, the hedgehog hibernates during winter in a burrow. Its temperature drops to a degree or so above freezing and its muscles set rigidly.

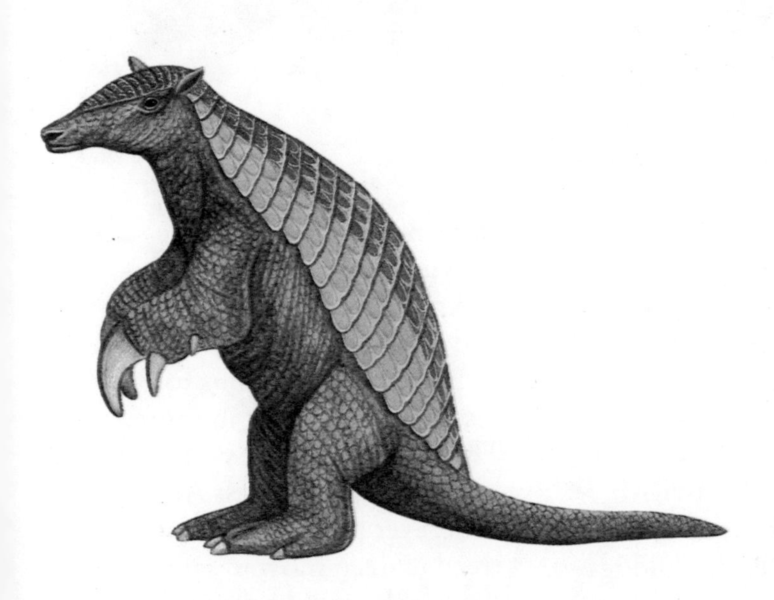

◀ **Armadillo** Many years ago a monster species of armadillo with a single-piece domed shell roamed the land. Today there are 20 species living in Central and South America. The largest is the **giant armadillo** – weighing up to 60 kg (132 lb) and 1.5 m (5 feet) from nose to tip of tail. Extremely tough, horn-like plates on head, back and sides provide excellent protection. Usually active at night, the giant armadillo digs for *termites* with its huge forefeet claws. Its diet also includes worms, spiders and snakes. When attacked by foxes or hawks, the giant armadillo delivers a fearsome blow with its foreleg. Sometimes it rolls itself up or digs down to safety.

## Dogs

Dogs were the first animals to be kept as pets. As long as 10,000 years ago there were two breeds of dog trained by people to hunt. Out of 37 species of dog (including wolves and foxes), all breeds of domestic dog belong to only one species. They are bred to do a certain job. Foxhounds and deerhounds are named after the animals they hunt. Terriers catch rats and dig out foxes. Retrievers bring back birds and rabbits that have been shot. All dogs have four toes on both front paws. The dog (male) is usually larger and stronger than the bitch (female). Up to 12 puppies are born in a single litter. For the first nine days they are blind and deaf. A dog's height is measured from the ground to the top of its shoulder.

◀ **Greyhound** This is one of the oldest breeds of dog. Greeks and Romans used them to chase hares; Egyptians raced them against each other. Greyhounds can reach speeds of 80 km/h (50 mph). They have semi-erect ears folded back. Their long legs, curved back and low-set tail are typical of running dogs. They grow to 75 cm (2½ feet) tall.

▲ **Mastiffs** stand over 75 cm (2½ feet) tall and weigh up to 75 kg (165 lb). There are many stories about mastiffs. The Roman army used them as guard dogs and for attacking the enemy. Later, mastiffs attacked knights on horseback. They wore coats of armour, had a flaming torch on their back and a spike sticking forward over their head.

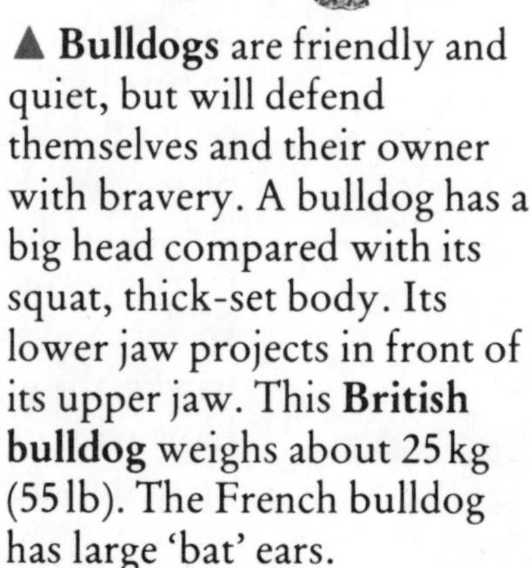

▲ **Bloodhounds** have a good sense of smell. The police use bloodhounds to follow the scent of people or animals. The loose skin on their heads hangs in folds. They have long, floppy ears. Bloodhounds are shy creatures. They grow to a height of 68 cm (27 inches).

▲ **Bulldogs** are friendly and quiet, but will defend themselves and their owner with bravery. A bulldog has a big head compared with its squat, thick-set body. Its lower jaw projects in front of its upper jaw. This **British bulldog** weighs about 25 kg (55 lb). The French bulldog has large 'bat' ears.

▲ **St Bernards** are the heaviest breed of dog – up to 91 kg (200 lb) with large head and big, drooping ears. St Bernards are famous for rescuing people lost in snow. They were used 200 years ago by monks in the Swiss Alps. Wherever a heavy dog could tread in deep snow in safety the monks could follow.

▲ **Chihuahuas** (pronounced chi-wa-wa) are the smallest dogs in the world – only 15–22 cm (6–9 inches) tall. They are also called 'pillow dogs' or 'ornament dogs'. Chihuahuas, bred in America from the Mexican hairless dog, are kept as house pets.

▲ **Pekinese** originate from China where it was once a pet of the Chinese court. These companion dogs walk with a rolling gait. Their long, straight coat touches the ground, hiding their legs. They have large eyes and wide, flat nostrils. Pekinese are 15–25 cm (6–10 inches) tall.

▲ **Terriers** are small, active dogs. They were used to chase vermin out of holes. The **Australian terrier**, a hunting terrier, stands 25 cm (10 inches) tall. British dog owners who emigrated to Australia interbred their terriers to create this new breed.

▲ **German Shepherd Dogs** are also known as Alsatians. They were bred as herding dogs. Today they are used by police and armed forces and as guide dogs for the blind. German Shepherd Dogs are 60 cm (2 feet) tall. They are intelligent and therefore easy to train. During the two world wars about 25,000 Alsatians lost their lives while working with the armed forces.

▲ **Spaniels** are gun dogs. This means they were bred to help the hunter. They were used to find and flush out game for the hunter to shoot. Other breeds were taught to fetch what was shot (called retrieving). The **British cocker spaniel** is one of several kinds of spaniel and is 38–41 cm (15–16 inches).

▲ **Great Danes** are known in Germany as the German Mastiff. In the 17th century Great Danes were used to hunt boars and stags in France, Germany and Denmark. They are excellent guard dogs. Their size is remarkable – up to 75 cm (2½ feet) tall and weighing over 54 kg (120 lb). When they stand on their hind legs, their front paws are as high as an adult person's shoulders.

▶ **Huskies** are hauling dogs specially bred to pull sleds over snow and ice in Greenland and north Canada. In the summer they help pull fishing boats ashore. These working dogs have great strength and endurance. They have a soft undercoat under their thick top coat. Huskies are suspicious of strangers and therefore make excellent guard dogs. They do not bark, but howl like a *wolf*. Husky dogs stand 60 cm (2 feet) tall and weigh up to 42 kg (92 lb); bitches are smaller – 55 cm (1¾ feet) tall.

◀ **Collies** are excellent working dogs, helping shepherds to herd sheep and cattle. There are now several breeds of collie. The **rough collie** originates from Scotland. Like the *husky*, the collie has a long outer coat and a short undercoat. Its semi-erect ears are mobile and expressive. It has intelligence, stamina and is faithful to its owner. When a rough collie walks, its front feet stay close together. This breed of collie dog stands about 60 cm (2 feet) high at the shoulder. Bitches are shorter.

▶ **African hunting dogs** live in the open plains, communicating by gestures, postures and a few calls. During much of the day they rest in shade and groom themselves. At dawn, dusk and on moonlit nights, packs approach herds of gazelle or zebra. They select their prey and chase it for up to 8 km (5 miles). When they catch the victim, the dogs gorge themselves on the meat. To bear their cubs, they make temporary dens in the burrows of other animals where usually seven cubs are born blind and stay until they can run with the pack.

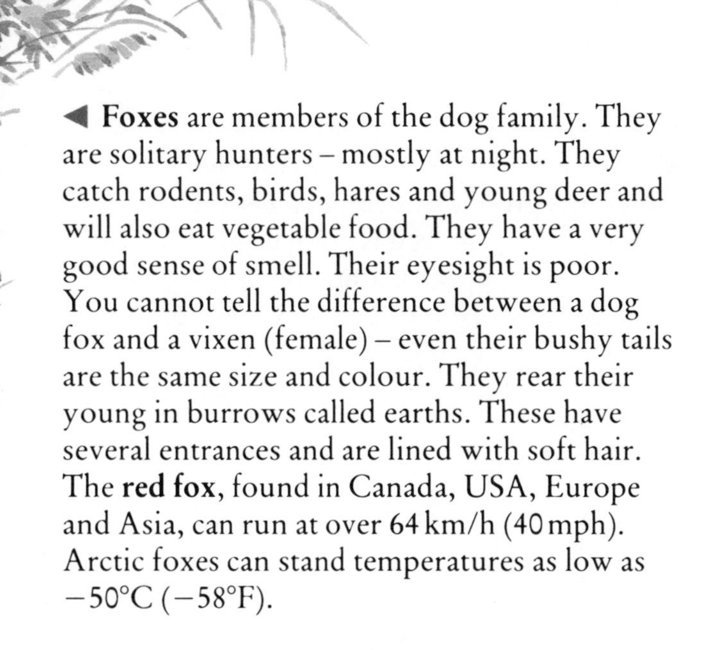

◀ **Foxes** are members of the dog family. They are solitary hunters – mostly at night. They catch rodents, birds, hares and young deer and will also eat vegetable food. They have a very good sense of smell. Their eyesight is poor. You cannot tell the difference between a dog fox and a vixen (female) – even their bushy tails are the same size and colour. They rear their young in burrows called earths. These have several entrances and are lined with soft hair. The **red fox**, found in Canada, USA, Europe and Asia, can run at over 64 km/h (40 mph). Arctic foxes can stand temperatures as low as −50°C (−58°F).

▶ **Dingos** are descended from dogs introduced into Australia many thousands of years ago. They live in family groups and gather in large packs to hunt. They fed on kangaroos until settlers started to kill kangaroos. Dingos then turned to sheep and rabbits for their food. They are good runners, able to sustain high speeds for long distances. Dogs and bitches look alike. Their bodies are about 1.5 m (5 feet) long and their tails are over 30 cm (1 foot). Four to eight puppies are born in each litter. The young stay with their parents for at least one year. Dingos' only enemies are humans, snakes and crocodiles.

◀ **Wolves** are the ancestors of all modern breeds of domestic dog. They are voracious eaters. A wolf can eat 10 kg (22 lb) at one meal. This is one-fifth of its body weight. In winter many species hunt in packs of 10 to 15. They hunt in the daytime for rodents, birds, sheep and goats. When on the trail, wolves lower their heads and raise the ruff on their shoulders. Wolves yip and howl to call to one another. The **maned wolf** is shy and does not hunt in packs. It is nocturnal and feeds mainly on small mammals. Its legs are very long. A maned wolf stands at 1.2 m (4 feet) tall.

► **Grizzly bear** Roaming North American forests, this huge animal can kill creatures bigger than itself with one blow of its paw. On its hind legs, a grizzly stands 2.5 m (8¼ feet) tall and weighs as much as 350 kg (770 lb). It is too slow to catch hoofed mammals and too heavy to climb trees. Its diet includes vegetation as well as fish and insects. In the cold tundra winter, the grizzly undergoes a period of torpor. This deep sleep is not true hibernation as the bear's temperature and breathing rate do not fall drastically and it often wakes up. Every two or three years a female grizzly has up to four young which are born blind, each weighing less than 700 g (1½ lb).

◄ **Polar bear** As tall as a *grizzly bear*, this great Arctic predator sometimes weighs in excess of 725 kg (1600 lb). It is well adapted for life in the snow and ice. Its creamy-white coat provides excellent camouflage and good insulation against the cold. Hair on the soles of its feet stop the cold penetrating and help the bear to grip the slippery surfaces. Surprisingly, polar bears can outrun *reindeer* over short distances. They are also good swimmers. They prey on seals, often waiting beside a seal's breathing hole. Fish, seabirds and arctic hares are also eaten. Blind and almost naked young, usually twins, are born in December in a den dug into a snowy slope.

► **Raccoon** A North American carnivore related to the *panda*, the raccoon is an agile climber and spends a lot of time in trees. It eats a wide variety of animals and birds: from rodents to cray fish, seeds to insects. A raccoon crushes its food with its strong back teeth. It will dabble a paw in water until it gets a 'bite' and then flips the victim out. Its bushy tail is ringed with black bands and its pointed face has a 'bandit' mask across the eyes. Three or four young are born in spring, weighing 70 g (2½ oz). Adult raccoons weigh 7 kg (15 lb) and are 1 m (3¼ feet) long.

◀ **Giant panda** This popular mammal from China is in danger of extinction in the wild due to the destruction of its main food source, the bamboo forests. Adult pandas are 1.5 m (5 feet) tall and consume huge amounts of food. A giant panda has five clawed toes on each foot and each forefoot has a small pad which acts as a thumb when grasping bamboo stems. It chews the bamboo with strong teeth and powerful jaw muscles. Giant pandas regularly climb trees for shelter. A single baby is born blind and helpless, weighing only about 140 g (5 oz). Its mother weighs as much as 115 kg (253 lb). The young panda grows rapidly and by eight weeks is 20 times its birth weight.

▶ **Polecats** are members of the same family as the *weasel, skunk, badger* and *otter.* They have supple bodies, short legs and a long tail. When alarmed, a polecat emits a nasty smell from a stink gland below the base of its tail. The **western polecat** lives in European forests. Males are larger than females, up to 60 cm (2 feet) long. These solitary creatures hunt at night for frogs, rodents, rabbits, birds' eggs and insects. They rarely climb trees, preferring to stay on the forest floor. The domestic ferret is probably descended from the polecat.

▼ **Badgers** These nocturnal creatures are so wary that they are rarely seen. The **Eurasian badger** occurs in forests across Europe and in Japan and south China. Across its head are two broad, black stripes. Family groups of badgers live in burrow systems called sets. Each set is used by successive generations. These playful animals feed on large quantities of earthworms and small animals, bulbs, fruit and nuts. Adult boars (males) are 80 cm (2⅔ feet) long and weigh 12 kg (27 lb).

◀ **Weasels** These small carnivores are related to stoats, minks and *badgers*. They are mainly nocturnal hunters who feed on small rodents and young rabbits, but they will attack animals twice their size. Weasels move so fast that it is difficult to follow them with the human eye. Despite this speed, weasels are caught and killed by hawks, owls and foxes. The **least weasel** or 'dwarf weasel' is the smallest carnivore. Its long, slim body enables it to follow mice into their burrows. They measure less than 30 cm (1 foot) from nose to tip of tail. Weasels have about five young in a litter.

▶ **Mongoose** These carnivores are members of the civet family. The **banded mongoose** lives in Africa, often near water. This good climber and swimmer is very bold in its search for food. Its diet includes insects, frogs, lizards, birds and fruit. It has been known to hurl eggs against stones to smash their shells. Banded mongooses live in family troops of up to 30. Their long bodies measure up to 75 cm (2½ feet) from snout to tail tip. The Indian mongoose is an expert snake killer.

◀ **Beavers** are master engineers. These rodents build complex dams from sticks and mud and they also erect dome-shaped lodges (homes). The lake behind the dam is used as a larder to store branches for winter food. They eat leaves and bark from aspen, birch and willow trees. These mammals close their nostrils when under water and can hold their breath for 15 minutes. Their dense coat is waterproof and they have webbed hind feet. A broad, flat tail, covered in scales, acts as an excellent oar. When in danger, a beaver will smack its tail on the water's surface and the sound is heard half a mile away. A male **American beaver** weighs up to 30 kg (66 lb) and its body is 1.3 m (4¼ feet) long.

▶ **Genet** This mammal, belonging to the same family as the civet, *mongoose* and *linsang*, has a slender body, narrow skull and short legs. **Small-spotted genets** are found in Africa, the Middle East and south-west Europe. These agile, graceful animals move on land with long tail held straight out behind. They climb well in trees. Genets spend the day sleeping and start hunting at dusk. They crouch almost flat when stalking, before pouncing on rodents, reptiles and insects. Small-spotted genets grow to 1 m (3¼ feet) from nose to tip of banded tail. Two or three young are born blind either in a tree, rock crevice or burrow abandoned by another animal.

◀ **Linsang** Like the *genet*, this slender and graceful mammal belongs to the civet family. The **banded linsang** is found in forests in Thailand, Malaysia, Sumatra and Borneo. It has four or five dark bands across its back and spots on its sides and legs. Its tail is long and ringed with dark bands. The banded linsang spends much of its life in trees where it climbs and jumps skilfully. It hunts at night for birds, small mammals, insects, lizards, frogs and birds' eggs. An adult banded linsang is 75 cm (2½ feet) from nose to tail. Little is known of their breeding habits. Their young are born in a hollow tree or burrow.

▶ **Hyenas** These relatives of the cat family have large, round ears. Packs of up to 30 hyenas hunt wildebeest and zebra across African plains at night. They can run at 64 km/h (40 mph). Hyenas communicate with sounds: whoops, growls, grunts, yelps and whines. The biggest and most aggressive is the **spotted hyena,** also known as laughing hyena because of its eerie yell when excited. An erect tail shows aggression, when pointing forward it displays excitement and when between its legs it means the hyena is afraid. The size of a large dog, spotted hyenas are up to 1.8 m (4½ feet) long.

**► Jaguars** These big cats are found in Central American forests and on the South American savannah. Their pattern of black spots is easy to distinguish from those of the *leopard* because they are arranged in a ring of four or five around a central spot. Black or albino (white) jaguars are known. Jaguars are normally solitary cats. Males and females only stay together for a few weeks when breeding. Up to four young are born in a den of vegetation either in a riverbank hole or among rocks. Females are aggressive when protecting their young from any intruder, including the father. Jaguars are excellent swimmers. They often climb trees to lie in wait for prey. They cannot sustain high speeds so depend on getting close to their prey without being seen. They prey on deer, tapirs, otters, turtles and fish. Adult males weigh up to 136 kg (300 lb). The biggest are 1.8 m (6 feet) long, with a 90 cm (3 foot) tail.

**◄ Tiger** Largest of the big cats, this powerful animal lives wild in India, Nepal, Bangladesh and Asia. Males and females look alike, although males have longer whiskers. The pattern of dark, vertical stripes varies from tiger to tiger. Its coat helps to camouflage this shy, solitary creature who hunts at night. Wild pigs, deer, cattle and small elephants are prey for the tiger. Adults kill about 30 times in a year. They eat 18–23 kg (40–50 lb) in one meal lasting up to two and a half hours. Tigers extend their sharp claws before seizing prey or when climbing trees. They walk with the claws retracted into a fold of skin so as not to blunt them. Two to four cubs are born and stay with their mother for several years. The biggest adult is up to 2.8 m (9¼ feet) long, with a 90 cm (3 foot) tail. All eight races of tiger are now endangered due to the destruction of their habitat and to being killed for sport or for their skins.

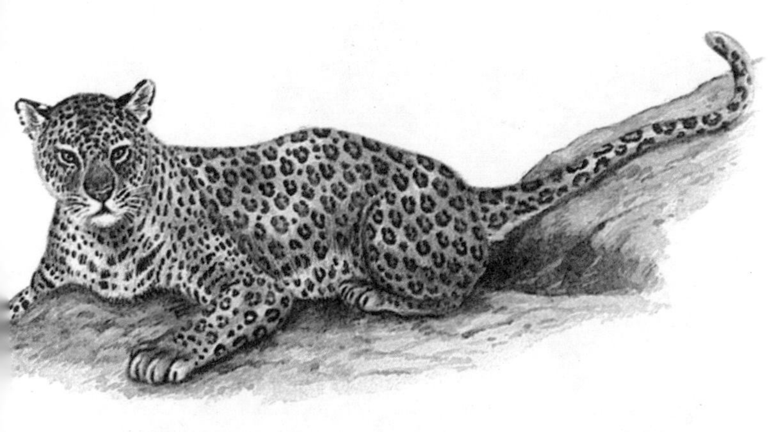

◀ **Leopard** Unique among big cats, the leopard carries large prey into a tree to devour it at leisure, safe from scavengers like the *hyena*. This solitary hunter has good eyesight and exceptionally acute hearing. The leopard preys on mammals up to the size of antelopes, young apes, birds, snakes and fish. It swims and climbs well, often lying on a branch to bask in the sun. An adult leopard's body is a little shorter than that of a lion but its tail is longer – 1.4 m (4½ feet). Leopards are found in Africa, Asia and the Middle East. Although their numbers are dwindling, leopards may prove to be the last of the big cats to survive.

▶ **Lion** 'King of the jungle' is an apt title for this splendid creature. It is feared by all animals on the plains in Africa, India and Asia. This big cat is very strong and the male has a heavy mane on its neck and shoulders. Its long tail is tipped with a tuft of hair concealing a claw-like spine. The male is larger than the female (lioness) and grows to 2 m (6½ feet), with a tail half as long. A group of lions is called a pride. Lionesses do the hunting, working together: some act as beaters driving the prey toward the others who are waiting in ambush. They prey on zebras, gazelles, giraffes and crocodiles among others. A lion may eat up to 34 kg (75 lb) of meat – one-fifth of its own weight – at a time. After a meal, lions sleep for hours. Up to six cubs are born in a litter, each weighing 1.5 kg (3 lb). Their coats are spotted, the spots becoming less obvious as they grow.

◀ **Snow leopard** This agile cat lives between the tree line and the snow line in the great mountain ranges of Asia. In winter, snow leopards migrate down to forests at about 2000 m (6500 ft). They stalk their prey such as wild sheep, goats and birds in the early morning or late evening. Snow leopards live alone, constantly roaming. Their furry coat is pale making it less easy to spot against the snow and hillside. Snow leopards are smaller than other leopards – only 1.5 m (5 feet) long, with a 90 cm (3 foot) tail.

► **Panther** This is an alternative name for the *leopard*, used especially for leopards whose coats are entirely black – the **black panthers**. These big cats are usually confined to humid Asian jungles. They are more athletic than other big cats such as the lion and tiger. Black panthers spend much of their time resting in trees. A black coat provides the panther with natural camouflage for its nocturnal hunting. It stalks its prey silently. Females hunt alone and hide any kill while they fetch their cubs. Usually two or three young are born in a rock crevice or a hole in a tree.

◄ **Cheetahs** are the fastest animals on land reaching speeds of 112 km/h (70 mph) over open country in Africa, Iran and Afghanistan. Also known as the 'hunting leopard', this big cat has a long tail which helps it balance during sprints. It stalks its prey and attacks with a rapid chase, knocking over its victim and killing it with a bite to the throat. It eats gazelles, jackals, hares and ground-dwelling birds (e.g. young ostriches). Occasionally several adults co-operate to chase and exhaust a zebra. Males grow to about 2.2 m (7 feet) from nose to tail. Cheetahs are becoming scarcer as they are hunted for their skins and sold as pets.

► **Bobcat** Like the *lynx*, but smaller and with less obvious ear-tufts, the bobcat is so-called for its short tail. Bobcats vary considerably in size: the smallest live in Mexico and the largest – up to 1 m (3¼ feet) – are found in the north of Canada. Bobcats live on the ground but climb trees, particularly when being chased. They are solitary animals, hunting at night in summer and during the day in winter. Their prey includes deer, goats, hares, rabbits, rodents, squirrels and grouse. After slowly stalking, they pounce on a victim's back, biting at its neck and tearing with their claws. Bobcats weigh 8 kg (18 lb) and often live for over 30 years.

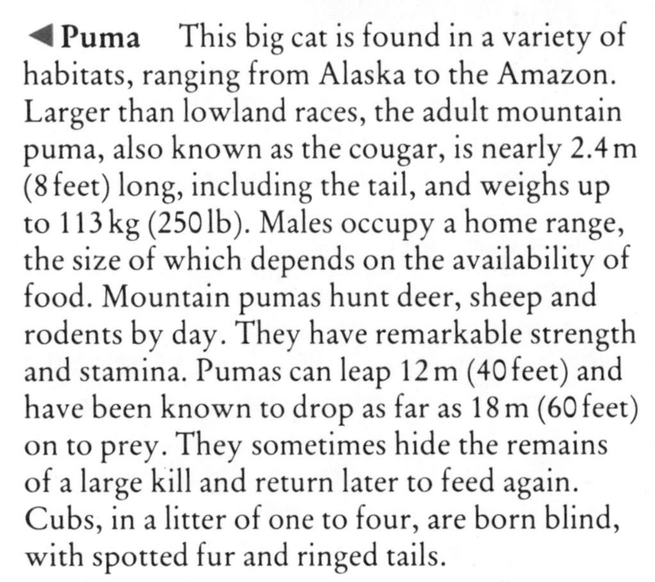

◀ **Puma** This big cat is found in a variety of habitats, ranging from Alaska to the Amazon. Larger than lowland races, the adult mountain puma, also known as the cougar, is nearly 2.4 m (8 feet) long, including the tail, and weighs up to 113 kg (250 lb). Males occupy a home range, the size of which depends on the availability of food. Mountain pumas hunt deer, sheep and rodents by day. They have remarkable strength and stamina. Pumas can leap 12 m (40 feet) and have been known to drop as far as 18 m (60 feet) on to prey. They sometimes hide the remains of a large kill and return later to feed again. Cubs, in a litter of one to four, are born blind, with spotted fur and ringed tails.

▶ **Ocelots** are very secretive cats found in Central and South American forests. Their characteristic dark markings are so variable that no two ocelots are alike. They sleep on a branch or in vegetation during the day and emerge at night to hunt for small mammals, such as wild pigs and *agoutis,* birds and snakes. An ocelot does not ambush its prey, but runs it down. The male emits loud, screeching calls when courting. Two to four young are born in a safe den in a hollow tree or in thick vegetation. Males grow to 1.7 m (5½ feet) from nose to tip of tail. Ocelots are now rare due to forest destruction and fur trapping.

◀ **Lynx** This wild cat has a stumpy, black-tipped tail, a short body and tufted ears and cheeks. Larger than a bobcat, it grows to 1.3 m (4¼ feet). The lynx is found in North America, Asia and northern Europe. This solitary, nocturnal hunter stalks its prey on the ground or lies in wait in vegetation. It can spring up to a branch 2.4 m (8 feet) above the ground or pounce with lightning speed on a feeding hare or a sitting grouse. Hares make up 70% of the North American lynx's diet. A lynx has broad, furry feet which enable it to travel over soft snow without sinking and to stalk silently.

## Domestic cats

Domestic cats make good pets, yet they retain a streak of independence. They will stalk and hunt birds and small animals. They are solitary animals; not like sheep, cattle or dogs, all of which prefer to be in groups. The Egyptians kept cats as pets nearly 4000 years ago. Their main ancestor was the African wild cat. The cat is an agile animal. Its long tail helps it to balance when walking along ledges. Its flexible back and strong muscles enable it to twist around in mid-air so that it always lands on its feet. When faced with an enemy, a cat arches its back and raises its fur to make it look bigger. A cat's sharp claws help it to scramble up rough surfaces. Its whiskers are sensitive to touch. An adult cat has 30 teeth. Its tongue is especially rough to help it keep its coat clean. Cats make many noises: they meow, hiss and purr. At mating time, tom cats (males) make a loud, wailing noise called caterwauling.

▲ **Tortoiseshell** cats are black and ginger. They are always female. Black cats and ginger cats sometimes produce a tortoiseshell kitten. If a tortoiseshell mates with a black or ginger tom, she may produce a tortoiseshell kitten.

1 **Siamese** cats originate from Siam (Thailand). They have thin legs, a long tail and large, pointed ears. They are cross-eyed. Their fur is a light colour with points (ears, face, feet and tail) a different colour. Their kittens are born with white fur; the colour on their points appears as they get older.

2 **Cornish Rex** cats have large ears, long legs, arched backs and strong hind legs. They can make high jumps and move with amazing speed. Rex cats have short, wavy fur and curly whiskers. The first known rex cat was seen in East Berlin in 1946. In 1950 the first Cornish Rex was born to a farm cat. This kitten was called Kallibunker.

3 **Russian Blue** cats originated from the port of Archangel in Russia. They are also known as Spanish Blue, Archangel and Maltese. They have wedge-shaped heads, tall ears and smooth, medium blue fur. Russian Blues are very affectionate towards each other and to their owners. They have quiet voices.

4 **Persian** cats come from Persia (now called Iran). They are known in some countries as Longhairs. These cats have a round head, small ears, short bodies and orange eyes. The Blue Persian is a very popular breed. The Chinchilla has long fur which is not completely white, for the tip of each hair is black. The effect is a coat of shimmering silver.

5 **Abyssinian** cats came to Britain from Abyssinia (now Ethiopia) in Africa during the 1860s. This is one of the oldest breeds. It is a very popular breed today, particularly in North America. This shorthaired cat has a ticked coat (each hair is banded with a darker colour) and ear tufts. Abyssinians look alert and lively.

6 **Maine Coon** are so called because they come from Maine, USA, and because the pattern of their coat looks like that of a *raccoon*. The males are very big and make good mousers. They weigh up to 18 kg (40 lb) and queens (females) never weigh more than 5 kg (11 lb). Their kittens are slow to develop, not reaching full size until they are four years old.

7 **Angora** cats were the first longhaired cats to be seen in Europe. They came from Angora (now called Ankara) in Turkey during the 16th century. Blue-eyed, white Angoras are often deaf. In Turkey, Angoras are named after their colour: sarman (red tabby), teku (silver tabby) and Ankara kedi (odd-eyed white).

8 **Burmese** cats do not come from Burma, but were bred in America. They have dark fur down the back which lightens toward the belly. Their large, golden eyes give Burmese cats an alert look. They have straight tails and small oval paws. Burmese are long-living cats – 16–18 years or more.

**Cats' eyes**

Cats can see in dim light. The pupils of the eyes enlarge, letting in as much light as possible. Behind each eye is a layer called tapetum which reflects any light. This causes a cat's eyes to glow in the dark. In bright sunshine the pupils close to a narrow, vertical slit. Cats have a wide variety of iris colours. Some white cats have blue eyes, others have orange eyes and a few have one blue, one orange eye.

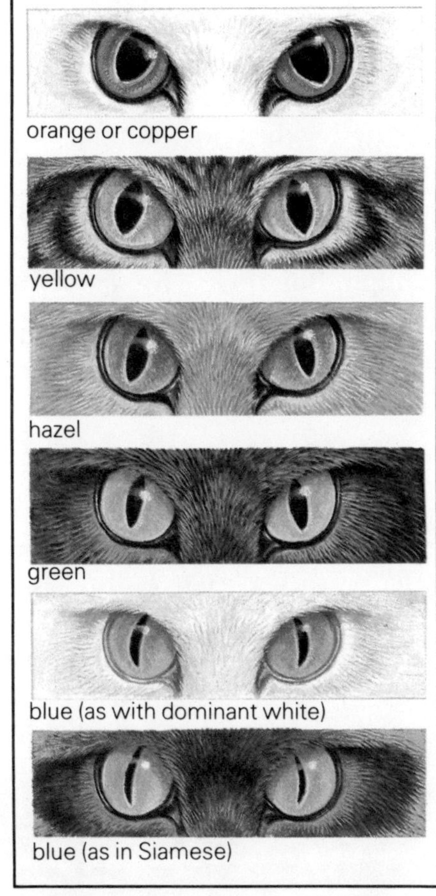


9 **Manx** cats come from the Isle of Man (between England and Ireland). These shorthaired cats have no tail, only a hollow where a tail would begin. There are many stories about how the Manx cat lost its tail. One story tells of Irish invaders stealing them to wear as plumes in their helmets.

◀ **Rhinoceros** Distantly related to the *horse*, this heavy mammal has extremely tough skin with only a few hairs. The two-horned African **black rhinoceros** is in fact grey but the colour varies depending on the mud it wallows in to cool off. Its pointed upper lip helps to pull tough leaves and buds off bushes. This shy creature is easily startled and may attack with its long horn, charging at 32 km/h (20 mph). A male black rhino is up to 3.6 m (11¾ feet) long, 1.75 m (5¾ feet) tall with a 70 cm (2¼ foot) tail.

▶ **Buffalo** This ruminant (cud-chewer) lives wherever there is enough grazing and plenty of water. Buffalo enjoy wallowing in mud to cool off and protect their hides (skins) from biting insects. African buffalo are aggressive fighters, especially when protecting their young, using their heavy horns and sharp hoofs. Crocodiles and lions sometimes succeed in killing young or sick buffalo. Herds of up to 100 feed mainly at night. Buffalo live for about 16 years and grow to a height of 1.5 m (5 feet).

▼ **Elephant** There are two species of this, the largest land mammal – the **Asiatic (Indian)** and African elephant. The Indian, at 3 m (10 feet) tall is smaller than the African at 3.5 m (11½ feet) tall. It also has smaller ears and shorter tusks and is used as a working animal. The elephant's trunk, an elongated nose, is used for smelling, gathering food and squirting water into its mouth or over itself. It has a diet of vegetation but when its six layers of teeth are worn down, the elephant faces starvation.

▶ **Hippopotamus** This name means river-horse. A large animal distantly related to *pigs*, the hippo spends most of the day submerged in African rivers. Often only its bulging eyes, ears and nostrils on top of its head are visible. A hippo's body is hairless, but it is protected by a pink, oily substance secreted from its skin. In the evening, hippos emerge to graze on land. They eat up to 20 kg (44 lbs) of grass in one night. Each foot has four toes with hoof-like nails. Male hippos yawn and bellow when angry. They can be 4.3 m (14 feet) long. The male **common hippopotamus** weighs about 4 tonnes and has a life span of about 50 years.

◀ **Hyrax** The hyrax is said to be the closest living relative of the *elephant*. The five species of hyrax are found in Africa and the Middle East. All have a cup on the sole of each foot which act as adhesive pads when climbing trees or rocks. Leopards and rock pythons are their main enemies. The **rock hyrax** feeds on leaves, plants and grass in the daytime. It makes a whistling sound when alarmed. A rock hyrax weighs up to 3 kg (6½ lb), is no more than 60 cm (2 foot) long and has no tail.

▼ **Warthogs** are members of the pig family. They get their name from the thick skin outgrowths beneath both eyes. These provide protection to the eyes while the warthog is grubbing for food. Males have long tusks: the upper ones are up to 60 cm (2 feet) long. Females give birth to 2–4 young. Many of the young do not survive. They either drown in the rainy season or are killed by lions, leopards or jackals. Adult warthogs grow to a length of 1.4 m (4½ feet). A warthog's tail becomes erect when it is angry.

▼ **Pigs** Domesticated pigs are descended from the wild boar. They can eat both plant and animal food. The Eurasian wild boar weighs up to 186 kg (410 lb) and the sow weighs 150 kg (330 lb). The boar has sparse, bristly hair and long tusks. Modern breeds of pig, like the **saddleback**, are the result of long and selective breeding. They look different from the wild boars: fatter, with shorter legs and less hair.

◀ **Antelopes** are members of the bovid family, which also includes goats, sheep and cattle. All bovids have split hooves and a four-chambered stomach. They are ruminants which means they chew the cud, bringing food up from the stomach and re-chewing it. The **roan antelope** has stout, ringed horns up to 1 m (3¼ feet) long and large, tufted ears. Small herds are preyed on by lions, leopards and hyenas. Male roan antelopes are aggressive and fight on their knees with vicious, backward sweeps of their horns.

▶ **Reindeer** These large deer are found in northern Europe and Asia. Nomadic Lapps migrated to the Arctic 1000 years ago with reindeer herds. They are dependent on reindeer for milk, meat, clothes, tents and tools. Reindeer have broad, outspread hooves which prevent them sinking in snow and muddy soil when running. They feed on lichens, moss, grass and leaves – any vegetation they can find. They migrate hundreds of miles between the tundra and winter feeding grounds further south. Their calves are born in May and June. Young reindeer run with the herd within a few hours of birth. Male reindeer weigh up to 315 kg (700 lb) and live for 12–15 years. They fight with other males using their branched antlers.

◀ **Gazelle** are related to the *antelope*. Enormous herds of **Thomson's gazelle** roam the Serengeti Plains in Africa. These small, graceful gazelle have a distinctive stripe along the side of their bodies. They are fast runners, capable of reaching speeds of 75 km/h (47 mph). Cheetahs, lions, leopards, hyenas and hunting dogs are their chief predators. The gerenuk (Waller's gazelle) feeds on acacia leaves. Although it is only 1 m (3¼ feet) tall, with its long neck it can reach foliage up to 2.4 m (8 feet) above the ground. The dorcas gazelle can go without water for 5–12 days. It gets body liquid from leaf sap.

► **Camels** Arabian camels (dromedaries) have a single hump. **Bactrian camels** have two humps. Camels are able to eat the thorniest desert plants. They store food as fat in their humps. A camel can drink 123 litres (27 gallons) of water in 10 minutes. Camels have become specially adapted for desert life: they close their nostrils during a sandstorm; toes connected by skin and tough, padded soles enable them to walk on hot sand; they travel for days without drinking; thick, coarse wool on upper surfaces is good insulation, but lack of fur elsewhere helps keep them cool; they do not sweat until their body temperature reaches 41°C (106°F). Bactrian camels live to a maximum of 50 years. They weigh up to 500 kg (1100 lb) and are 1.8 m (6 feet) tall at the shoulder.

◄ **Giraffe** This tallest of all living animals lives in Africa. It uses its long tongue and upper lip to eat acacia leaves. Giraffes do not move about much. They keep three feet in contact with the ground at all times. They can give a deadly kick to attacking lions. On their heads are two to five stubby horns covered in skin. A giraffe has to spread its front legs wide apart to drink. At night it rests standing. A young giraffe weighs 53 kg (117 lb) and is 1.8 m (6 feet) tall at birth. They grow for 10 years, when they reach their full height of 5.3 m (17½ feet). They live up to 35 years.

1 **Goats** Like *sheep* and *cattle*, goats have hoofs and a four-chambered stomach. Both male and female goats have horns which curve upwards and backwards. Male goats have beards and a strong odour. **Domestic goats** stand 1 m (3¼ feet) at the shoulder and weigh up to 118 kg (260 lb). The oldest recorded age for a goat was 20 years 9 months. Goats do not sleep; they have periods of drowsiness. They feed mainly on leaves and twigs, but sometimes eat paper and cloth. The mountain goat of North America is really a goat-antelope. It lives so high up in the Rocky Mountains that no predators can reach it.

2 **Cattle** There are two species of large, hollow-horned ruminants called cattle. They are the western cattle and the zebu of Asia and Africa. Western cattle are descended from aurochs, the wild ox of Europe and Asia which became extinct in 1627. All cattle today are domesticated. Farmers divide cattle breeds into three types: beef, dairy and dual-purpose. Beef cattle are intended primarily for meat production. The average milk yield from a **British Friesian cow** is 20 litres (4.4 gallons) in one day. Some bulls (males) weigh up to 1000 kg (2200 lb) by the age of two and a half.

3 **Sheep** were the earliest ruminants to be domesticated, probably as early as 12,000 years ago. Today there are about 450 breeds of domestic sheep. Some are hornless, others have two, four or eight horns; some are tall, others short; some are white-faced, others black-faced; some have dense, woolly fleeces, others are without wool. Lambs are suckled by their mothers for about six weeks. The **merino** is bred for its high quality wool. It can be horned or hornless and it is found chiefly in Australia, North America and Spain. Sheep are basically plant-eaters but they will also eat snails.

1 **Mongolian wild horse** This is one of the few breeds of wild horse left in the world. It is also called Przewalski's horse after a Russian explorer who found the first herd in Mongolia, central Asia, in 1881. There are not many Mongolian wild horses left today. They are distinguished by their long-haired black tail and erect mane. Their summer coat has a dark stripe down the middle of the back. They stand 130 cm (4⅓ feet) at the shoulder. These wild horses are the most primitive living members of the horse family and living ancestors of the domestic horse.

2 **Donkey** This descendant of the wild ass of Africa is a domesticated member of the horse family. Until the end of the eighteenth century, donkeys were called asses. By this time these beasts of burden were to be found over a large part of the world, carrying riders or pulling ploughs. Breeds vary in colour from nearly white to nearly black. The mahratta breed of Sri Lanka and Pakistan is the smallest – 75 cm (2½ feet) high at the shoulder. The largest is the Poitou donkey of France – up to 1.6 m (5¼ feet) high. Donkeys have lived up to 24 years of age in zoos.

3 **Zebra** An adult zebra weighs 317 kg (700 lb), stands 1.2 m (4 feet) at the shoulder and lives for up to 62 years. A zebra can reach a speed of 64 km/h (40 mph). Within each herd in the wild there are groups consisting of a stallion (male), six mares and their foals. When alarmed, a zebra stallion emits a braying call. This sets the whole herd running. The stallion stays at the back of the family group and kicks out at attacking lions and hyenas. The unusual features of **Grevy's zebra** are its large rounded ears, the narrow space between its stripes and its upright mane.

4 **Mule** This half donkey half horse is another beast of burden. It is a strong pack animal. Like zebras and donkeys, the mule has a foot which has evolved into a single hoof. Its teeth are adapted for cropping and grinding grass and its skull is elongate to allow room for large cheek teeth. The mule defends itself by kicking out backwards with its hind legs. Like all members of the horse family it is a fast runner. Most adult mules are about the same height at the shoulder as a zebra, but have a longer neck. The record age for a mule is 37 years.

## Horses and ponies

The story of the horse began 55 million years ago when a strange little animal called Eohippus (Dawn Horse) lived in swampy forests. Twenty million years later, the Dawn Horse's successor, Mesohippus, moved out of the forest and changed its diet from leaves to grass. These animals had to move fast to escape danger; so early horses, like Merychippus, had long legs and ran on their middle toes. Five million years ago, Pliohippus was the first horse to have only one toe on each foot protected by a hard nail, the hoof. The average height of a modern horse (Equus) is 1.6 m (5¼ feet). A pony is a small horse, no higher than 1.4 m (4⅔ feet).

Horses and ponies have been domesticated for over four thousand years. At first they were used to pull loads and carry goods; later their human owners learned to ride them. They are divided into types and breeds depending on their size and the work they do. There are about 300 breeds in the world today. Horses and ponies are measured in 'hands' from the ground to the top of the withers. A 'hand' is equal to 10 cm (4 inches).

**Eohippus**, or Dawn Horse, existed for 15 million years until 40 million years ago. It was only 20 cm (8 inches) tall, ate leaves and had four toes on its front feet and three toes on its hind feet.

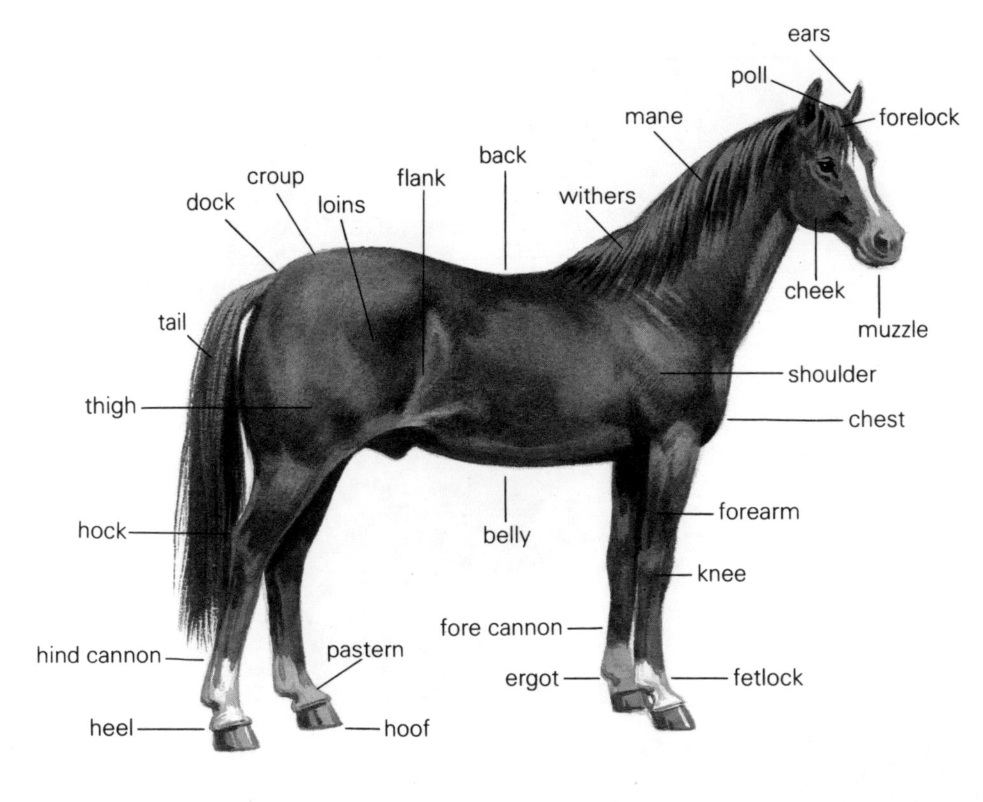


Whatever their type and breed, horses have the same points (different parts with special names).

◄ **Arab** This is the oldest and purest breed of modern horse. Arab horses are known to have lived in the deserts of Arabia 7000 years ago. The Prophet Mohammed used Arab horses for his successful cavalry in the 7th century. Huge numbers of these desert-bred horses went on campaigns through North Africa and into Spain and France. Nearly all modern horse breeds can be traced back to Arab forefathers. Today Arab horses are bred in large numbers all over the world. Adults are 14–15 hands high. Their colours are chestnut, bay or grey.

1 **Thoroughbred** This breed was developed specially for horse racing. It is also called 'English Racehorse'. Thoroughbreds are the fastest horses over short distances: some reach speeds of 70 km/h (43 mph). They are all descended from just three sires (fathers) called Byerley Turk, Darley Arabian and Godolphin Barb, imported into England in the late 17th century. The average height of a Thoroughbred is 16 hands.

2 **American Quarter Horse** This breed is a star at rodeos and in the show ring. It is one of the oldest and most popular breeds in America. The first horse races in North America were run along 'race paths' a quarter of a mile (0.4 km) long. This is why the horses which ran in the races were called 'Quarter Horses'. The American Quarter Horse is 15–16 hands tall and is usually a chestnut colour.

3 **Shire** This is the tallest horse in the world – it can be 18 hands. It is a draught horse, capable of pulling loads of 5 tonnes. A draught horse is the world's second strongest animal (after the elephant). The Shire is named after the counties called shires in central England. These horses are slow but steady workers. They all have white on the feet and legs.

4 **Haflinger** This small but sturdy pony from Austria stands 14 hands high. Its colour is often palomino (golden with flaxen mane and tail). Haflingers were used as pack ponies by the German army. They were also used in agriculture and forestry. Today they are popular with tourists for pony trekking.

5 **Shetland** One of the world's smallest breeds, this strong pony originates from the Shetland islands off northern Scotland. They were used down coalmines in the 19th century. Shetland ponies have round feet and grow a double coat of fur in winter. They are rarely more than 10 hands high.

# Rodents

Rodents are small mammals with a single pair of sharp teeth at the front of their upper and lower jaws. These teeth continually grow to replace the parts worn away by gnawing. They eat mainly plant material and are often regarded as pests.

► **Agouti** This shy South American rodent digs burrows in river banks, under trees or stones. It follows the same route to its feeding grounds. It hides its food – nuts, fruit and leaves – in several different places. It sometimes peels its food before eating it. Like all rodents, an agouti has a long, rounded body with hind legs longer than front legs. An adult is up to 65 cm (26 inches) long.

◄ **Squirrels** are always twitching their noses and whiskers to sense danger. They use their bushy tails for signalling and to balance as they run along branches. There are 246 species of squirrel. They live in European and Asian evergreen forests, feeding on conifer cones, fruit and fungi. Their young are born in dreys (tree nests made of twigs, moss, fur and feathers). Squirrels lay dormant in winter, waking every few days to eat. **Red squirrels** grow to 25 cm (10 inches) and have a tail nearly as long.

▼ **Rats** are found worldwide. Their success is largely due to their rapid rate of breeding: 5–10 young are born every six weeks throughout the year. These rodents are omnivores and will attack animals larger than themselves when food stores are empty. **House rats**, also known as black or ship rats, carry diseases such as bubonic plague, typhus and rabies. Over the past 1000 years more people have died from these diseases than in all wars. Adult house rats are 25 cm (10 inches) long, with a tail of equal length.

◀ **Moonrats** are members of the hedgehog family found in forests and mangrove swamps in South-East Asia. This scruffy creature is shy and nervous. Its long snout is bristling with whiskers and its tail is naked and scaly. A moonrat emerges from hollow logs or between tree roots at dusk to search for insects, worms, fish and crabs. It defends itself by producing a foul smell which repels most attackers. Little is known of its breeding habits, although two young are probably born each year. The body of an adult male is 45 cm (1½ feet) long, with a tail half that length.

▶**Mole** All 20 species of mole dig very long burrows in soft soil. A **European mole** can dig up to 20 m (66 feet) in a day. Its front feet and forelegs have evolved to act as both pickaxe and shovel. Its tiny eyes are covered with hairy skin. Bristles on its face feel vibrations in the soil. When a mole reverses, it holds its stumpy tail erect and hairs on it provide warning of any danger. European moles are 20 cm (8 inches) long and weigh up to 120 g (4½ oz). They feed mainly on earthworms. A litter of up to seven young are born in a leaf-lined nest underground.

▼ **Shrews** live among debris on forest floors. More than 200 species are found throughout the world, except in Australasia and the West Indies. Shrews communicate by shrill squeals which are so high-pitched that we cannot hear them. They rely on smell and hearing when hunting insects. Shrews eat their own body weight in food each day. When young **long-tailed shrews** (found in Sri Lanka) leave the nest, each one grips the tail of the one in front in its mouth. Led by the mother, this 'caravan' goes in search of food. Pygmy shrews – 8 cm (3 inches) from nose to tail – are the smallest living land mammals.

▲ **Pocket mouse** There are about 70 species of pocket mouse and kangaroo rat in North, Central and South America. They have deep cheek pouches lined with fur. Seeds are taken back to burrows in these cheek pouches. The **Californian pocket mouse** digs deep burrow systems in sandy soil. They seldom drink; their bodies are adapted to survive without water. Up to seven young are born in each litter. Adults measure 36 cm (14 inches) from nose to tail.

▲ **Hamsters** These rodents store food in cheek pouches. They have thick-set bodies with short legs and a short tail. **Golden hamsters** come from the Middle East. They have the shortest pregnancy of all mammals: 15–16 days. They are nocturnal creatures that eat both plants and animals. Adults live alone in burrows and measure up to 18 cm (7 inches) from head to tail. Golden hamsters have become popular pets.

▲ **Lemmings** are related to *voles.* **Norway lemmings** live 760–910 m (2500–3000 feet) above sea level on Scandinavian mountains. They feed on grasses, shrubs and mosses. Their enemies are foxes and birds of prey. These fearless, agile rodents have many babies: 4–10 young in three litters each year. Every few years the lemming population becomes too large, so some lemmings migrate. Many drown in rivers, lakes and the sea. Norway lemmings measure 15 cm (6 inches) and weigh up to 112 g (4 oz). Their life span is only two years.

▲ **Voles** have small ears, short tails and blunt faces. They are related to *lemmings.* Their enemies are foxes and birds of prey. The European **bank vole** eats berries, grasses, buds, shoots and some insects. It is active day and night, with several rest periods. Bank voles are good swimmers and climbers. Their nests are usually under logs or among tree-roots. They are smaller than other species – 15 cm (6 inches) long, including the tail.

▲ **Dormouse** The **fat dormouse** lives in European and Asian forests. It is the largest of the 14 species of dormouse. This agile tree-climber has rough pads on its feet to help it grip. It nests in the fork of branches, rock cavities or abandoned rabbit burrows. Fat dormice feed on seeds, fruit, buds, leaves and occasionally on insects. They build up reserves of body fat before hibernating between October and May. Adult fat dormice measure up to 33 cm (13 inches) from nose to tip of their bushy tail. They weigh about 114 g (4 oz) and live for a maximum of 12 years.

▲ **African climbing mouse** All 21 species of climbing mouse live in Africa. This little rodent has a prehensile tail which acts as a fifth limb when climbing plant stems. It feeds at night on berries, fruit, seeds and sometimes small lizards. There are incidents of African climbing mice raiding the nests of weaver finches and making their own nest there. The African climbing mouse is only 10 cm (4 inches) long and its tail is slightly longer. They breed throughout the year, with litters of three to five young.

▲ **Rabbits** live anywhere that offers grazing and soil in which they can burrow. When frightened, a rabbit thumps its hind feet on the ground and raises its tail to warn others. Rabbits live together in groups of burrows called a warren. Does (females) look after their young in nests of hay and fur. Each doe has an average of 10 babies a year. The North American **brush rabbit** is shy. It feeds on almost anything. Bucks (males) measure 36 cm (14 inches) from nose to tail.

▲**Hares** are fast runners with strong hind legs and long, narrow ears. They live in shallow nests called forms. **Brown hares** are found in grasslands in all continents. They feed on grass, cereals, clover, root vegetables and tree bark. The brown hare's speckled appearance is due to black-tipped hairs. 1–5 leverets (baby hares) are born in a form, fully furred and with their eyes open. Adults weigh up to 6.5 kg (14 lb). They can reach speeds of 75 km/h (45 mph).

**▶Sea-lion** These mammals are well adapted to living on land but spend most of their time in the sea. Unlike seals, they have external ears. They are quite tame except when defending their breeding territory or a pup. Large numbers haul themselves out of the water at traditional breeding sites called rookeries. They tuck their hind flippers forward to move over land. The largest species is the **Steller sea-lion**. Males – 3 m (9½ feet) long – are considerably larger than females. They have big heads and thick necks. In the Pacific Ocean these sea-lions dive to 180 m (600 feet) in search of fish, squid and octopus.

**◀Otter** This relative of the *skunk* is a master swimmer, well adapted for life in water. Its body is streamlined and all four feet are webbed. A thick, muscular tail helps propulsion and nostrils and ears can be closed when under water. Short, dense fur keeps the otter dry by trapping a layer of air around the body. The **giant otter** of South America is an endangered species. Unlike other species, giant otters usually travel in groups. They feed during the day on fish, eggs and mammals. One or two young are born in a den in river banks or under tree roots. Giant otters grow to 2.2 m (7 feet) from nose to tip of flattened tail.

**▼Walrus** Found mainly in the Arctic Ocean, walruses feed on molluscs, crustaceans and starfish. They eat as much as 60 kg (132 lb) in a day. The male is a huge mammal with two tusks. By swimming upside down and striking upward with its tusks, a walrus can kill a seal or open up a breathing hole in the ice. It also uses its tusks to haul itself on to ice floes. Males grow to 3.5 m (11½ feet) and weigh about 1360 kg (3000 lb). Females are smaller and have shorter tusks. A single pup is suckled by its mother for two years. A walrus has no external ears, hairless skin and small, bloodshot eyes.

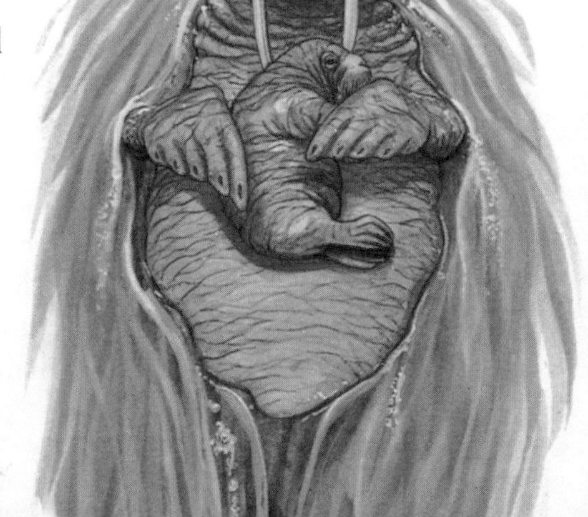

◀ **Elephant seal** When angry bull seals rear up, they rise to over twice the height of a human. These males fight one another for dominance. They give loud vocal threats to rivals – their enlarged snout acting as a resonating chamber. Male **northern elephant seals** are almost 6 m (20 feet) long and may weigh 2700 kg (6000 lb). They have a thick layer of blubber (fat) under their skin. Elephant seals eat fish and squid. The much smaller female gives birth to a pup on islands off the Pacific coast of America. She defends her pup until it is ready to take to the sea.

▶ **Monk seal** The only seal species found in tropical waters, monk seals are becoming rare for several reasons: their breeding grounds are disturbed by people; they die from pollution or are accidentally entangled in fishing nets. In 1985 the estimated population of the **Mediterranean monk seal** was 500. Males grow to a length of 2.7 m (8¾ feet). A seal is a skilful swimmer and diver. When it dives for food its heart rate slows from 120 to four beats a minute. This allows it to stay submerged for long periods. Unlike sea-lions, seals do not have external ears. Most seals cannot turn their hind flippers forward, so movement on land is restricted and these seals drag themselves along by their fore-flippers.

◀ **Grey seal** This seal, also called the Atlantic seal, is found in the North Atlantic Ocean. Males are up to 2.3 m (7½ feet) long, weigh 213 kg (470 lb) and live as long as 40 years. Cows (females) are half the weight of bulls and live for only 30 years. Grey seals travel far from their rookeries but stay mostly in coastal waters. They feed on fish, crustaceans, squid and octopus. Grey seals can turn their hind flippers forward to move on land. Females arrive at their breeding grounds first and give birth to a single pup before the male appears. The pup puts on 22.5 kg (50 lb) in weight in two weeks.

▲ **Humpback whale** Found in all oceans, this huge mammal grows to 19 m (62 feet). It is also known as the fin whale because of its 5 m (16 feet) long flippers which have uneven edges. There are, on average, 22 grooves under its curved lower jaw. Family groups of three or four feed in polar regions in summer and migrate to tropical breeding seas for the winter. A single calf is suckled by its mother for almost a year before it starts to feed on crustaceans, fish and squid. Humpbacks sing for hours on end. After a dive, whales surface and blow air out of a blowhole in their head.

▼ **Blue whale** The largest species of mammal that has ever lived, the blue whale can weigh over 146 tonnes (160 US tons). This is equal to the weight of 30 elephants. A blue whale is up to 32 m (105 feet) long. Its heart beat slows to 5–10 beats a minute when swimming slowly. During the summer, blue whales feed on krill (small shrimps) in polar waters. The stomach contents of one whale weighed 425 kg (936 lb). In autumn, blue whales migrate towards the Equator. Due to whaling, they are now in danger of extinction.

▼ **Sperm whale** This is the largest of the toothed whales and is found in warm oceans. It has a huge head and several fleshy humps on its back. There are teeth only in its lower jaw. The sperm whale can dive to over 900 m (3000 feet) and remain in the depths while searching for prey. It has a kind of sonar system to find large squid, lobsters and other marine creatures. Fights with squids often leave the sperm whale scarred. A bull sperm whale grows to 20 m (65 feet) and weighs up to 75 tonnes (83 US tons).

▼ **Right whales** were named by whalers who regarded it as the 'right' whale to hunt. As a slow swimmer it was easy to catch and when killed it floated and so was towed behind the ship to attract other whales. There are now very few right whales left. Like *blue whales*, they swim along with their mouths open, sieving large amounts of krill (small shrimps). They do not have teeth, but trap food behind bony plates in their jaws called baleen. *Barnacles* stick to the flesh around the whale's mouth and feed on scraps that the mammal spits out. Right whales are 20 m (65 feet) long.

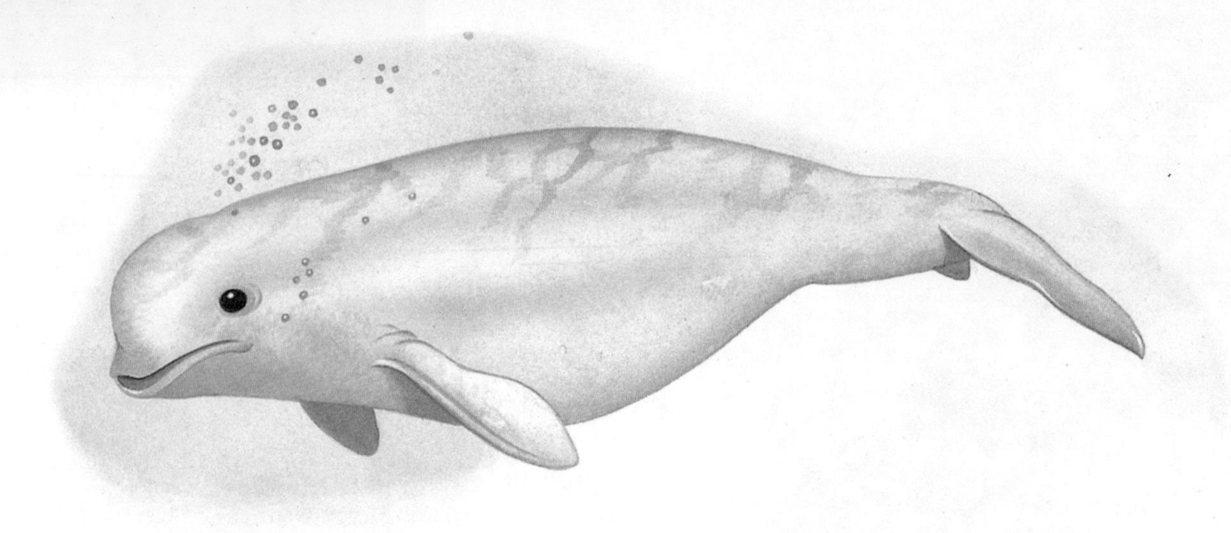

▲ **Beluga** One of two species of white whale living in shallow seas in the Arctic, the beluga has a round, plump body. At birth a beluga calf's smooth skin is a dark brownish-red colour. This changes to blue-grey as the calf grows. The skin gradually becomes paler and by the age of six it is creamy-white. Adult belugas grow to 6 m (20 feet). Hundreds of them migrate south in winter. They feed on crustaceans, sand-eels, cod and other fish. Belugas swim under pack-ice and break through the ice to breathe. Their songs can be heard above water and they were known as 'sea canaries' by whalers.

▼ **Narwhal** The male of this species of white whale has an extraordinary spiral tusk which is fragile and easily snapped. It is actually a tooth that grows out of a hole in the narwhal's upper lip. It can be as long as 2.7 m (9 feet). It is not known what this tusk is for, although it is possibly an attraction to females. A narwhal calf has dark blue skin at birth. As it matures the colour changes to the mottled brown of the adult. Narwhals feed on squid, crabs, shrimps and fish. Large herds may gather when migrating southward for the winter months. Although narwhals make clicking and other vocal sounds, no one is sure whether these are used to locate prey or simply for communication.

◀ **Dolphin** The 32 species of dolphin form the largest family of toothed whales. Dolphins swim fast and can reach speeds of 60 km/h (37 mph) when riding waves. They have a slender body and a long snout. Inside the bulging forehead is a pad of fat shaped like a lens which helps to focus sonar beams. Up to 1000 clicks per second are made by the **bottlenose dolphin**. This highly intelligent mammal judges distances with great precision. It eats up to 15 kg (33 lb) of fish each day and grows to a length of 4.2 m (14 feet).

▶ **Porpoise** This small whale is rarely over 2.2 m (7 feet) long. Most of the six species have dorsal fins. They live in coastal waters in the northern hemisphere, often in estuaries. The **common porpoise** dives for up to six minutes and locates fish, particularly herring and mackerel, by sonar. Up to 15 porpoises in a group called a 'school' communicate with each other by a series of clicks. Male and female common porpoises have a long courtship, caressing one another as they swim. As with all whales, the young calf is born tail first in the water so that it does not drown. The mother takes the calf to the surface to breathe air.

◀ **Killer whale** Despite its name, this strong mammal is really a dolphin. It has characteristic black and white markings and a rounded head. Larger than other species, the adult male is 9.7 m (32 feet) long and its dorsal fin is about 2 m (6½ feet) high. Killer whales live in the colder seas worldwide. They are keen predators and have 40–50 teeth. Family groups co-operate in hunting fish, squid, penguin, seal and even whale. The killer whale does not hunt humans. Unlike other dolphins, this species makes different click sounds that last for 25 seconds. A killer whale calf is nearly 2.2 m (7 feet) long when born.

◀ **Flying squirrel** This rodent takes flight when an owl or other predator appears. It stretches out all four limbs as it jumps off a branch opening out the skin stretching from its wrists to its ankles. This 'parachute' allows the squirrel to glide to the next tree. **Northern flying squirrels** live in forests in Canada and the USA. These shy, nocturnal creatures store nuts and berries in hollow trees ready for the winter. The female cares for her young for about 10 weeks. This level of development is necessary before the young can attempt gliding. Adult northern flying squirrels are 45 cm (18 inches) long, including the bushy tail.

▶ **Sugar gliders** are marsupials found in trees in Australia and New Guinea. They look like *northern flying squirrels.* They glide from tree to tree and can cover distances of up to 55 m (180 feet) at a time. A sugar glider flips its body upward to land on a tree trunk. It often returns to the same tree again and again. It feeds on insects and nectar. Groups of up to 20 sugar gliders live in a nest of leaves in holes in eucalyptus trees. A sugar glider carries nesting material in its curled tail. Adults have a 15 cm (6 inch) body with a longer tail.

◀ **Flying lemurs**, or colugos, feed at night on leaves, fruit and shoots. During the day they hang by their curved claws from branches, with their heads up. They 'fly' with the aid of a fold of skin stretching from neck to ankles and to tip of tail. The **Philippine flying lemur** can glide for up to 135 m (150 yards). Its enemy is the Philippine eagle. Females carry their young until they are too large and too heavy. From nose to tail an adult flying lemur measures 70 cm (2¼ feet).

## Bats

Bats are the only mammals that can actually fly. Their wings consist of thin, elastic skin stretched between four very long fingers on each forelimb and its ankles. There is a hooked claw on the movable thumb. Around the world there are 590 species of bat, divided into two main groups; fruit-eaters and insect-eaters. Most bats are nocturnal hunters. They make supersonic squeaks that bounce back from solid objects, including the food they hunt, and are picked up by their sensitive ears.

◀ **Vampire bat** This South American mammal feeds solely on other animals' blood. Four sharp teeth pierce the victim's skin and it licks up the blood with its long tongue. Vampire bats feed for half an hour each night. They transmit various diseases in their saliva. An adult vampire bat is 9 cm (3½ inches) long and has a wingspan of 18 cm (7 inches).

▶ **Horseshoe bats** have an extra fleshy structure surrounding the nose. This acts as a megaphone when they 'shout' their squeaks through their nostrils. When roosting, the horseshoe bat wraps its wings round its body. The **greater horseshoe bat** is found in Europe, Asia and North Africa. It is slow and fluttering in flight and feeds by swooping down on beetles. Adults are 12.5 cm (5 inches) long, with a small tail and a wingspan of 34 cm (15 inches). Males and females look alike.

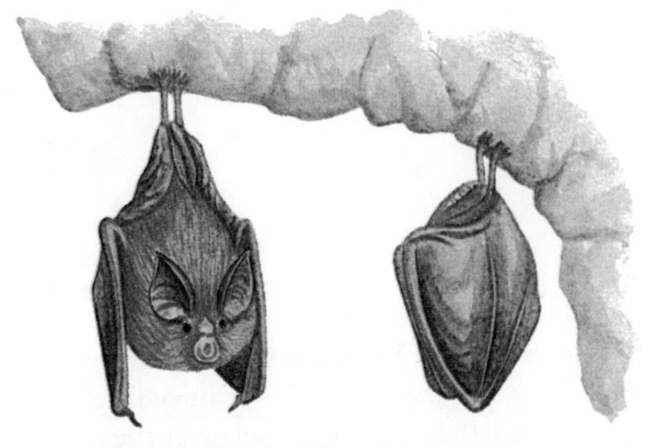

◀ **Pipistrelle bat** This nocturnal insect-eater hibernates for five or six months during winter in mines, caves or cellars. In summer it hides behind window shutters, in crevices and nest-boxes during the day. The most common European bat is the **common pipistrelle**. Each pair has two young every year. They grow to a length of 7.5 cm (3 inches), including the tail. Their wingspan is 25 cm (10 inches).

► **Long-eared bats** belong to the evening bat family. They have very long ears. **Common long-eared bats** are found in woodland areas in northern Europe and Asia. They hunt for moths, mosquitoes and flies at dusk, often dive-bombing their prey. In summer they live in lofts, trees and birds' nestboxes. Between October and March they hibernate in caves, mines and cellars. They live for 12–15 years and weigh only 7 g (¼ oz). Their wingspan is 28 cm (11 inches).

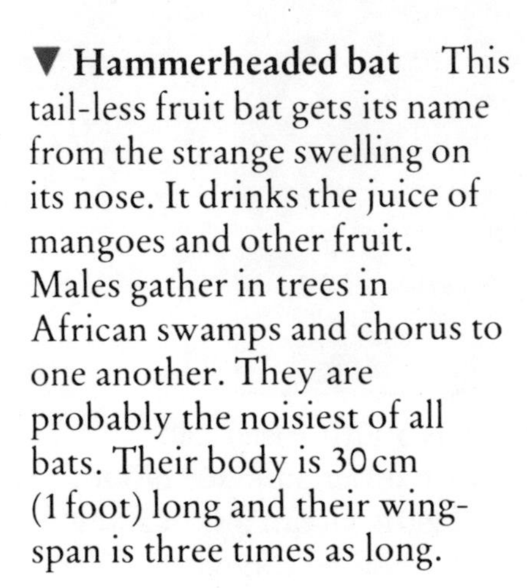

▼ **Hammerheaded bat** This tail-less fruit bat gets its name from the strange swelling on its nose. It drinks the juice of mangoes and other fruit. Males gather in trees in African swamps and chorus to one another. They are probably the noisiest of all bats. Their body is 30 cm (1 foot) long and their wing-span is three times as long.

▼ **Leaf-nosed bats** There are 140 species of leaf-nosed bat, so called because of a flap of tissue above their nostrils. They are efficient food processors: a meal of fruit passes through their gut in 15 minutes. Seeds are dispersed in their droppings. The **Jamaican fruit-eating bat** is found in the West Indies and in some South American countries. Its body is up to 9 cm (3½ inches) long, with a wingspan of 26 cm (10½ inches).

▼ **False vampires** These bats have long canine teeth, but are not blood-suckers. The **Australian false vampire** feeds on lizards, mice and other bats. It flops down on its prey, trapping it in its wings, before biting the back of its victim's neck. This bat is also known as the 'ghost bat' because of its pale colour. Its body is 15 cm (6 inches) long but its wingspan is 60 cm (2 feet).

▼ **Fruit bats** are also known as 'flying foxes'. Males and females look alike. They have large eyes and a keen sense of smell. They have a strong claw at the end of each thumb. Using these claws and their hind feet, they make their way along branches. **Greater fruit bats** of Asia have the largest wingspan of any bat – 1.5 m (5 feet). They use peg-like teeth to crush fruit. They swallow the juice and spit out the seeds and flesh.

# REPTILES and AMPHIBIANS

Amphibians evolved between 375 and 350 million years ago from now extinct lung-breathing fish which emerged from the water and moved on the land using strong fins. Amphibians developed legs and eyelids to become the first vertebrates to live on land but they were, and still are, dependent on water for breeding. It was the reptiles which became the first group of vertebrates to become truly adapted to life on land.

Reptiles have scaly, waterproof skin and most lay their eggs in soil. The eggs have tough shells to protect them from damage and water loss. Young reptiles are miniature versions of their parents and breathe using lungs. Like amphibians, reptiles are 'cold-blooded' animals as their body temperature is close to that of the air or water around them. They gain heat and energy by basking in the sun. There are four groups of reptile: turtles and tortoises; crocodiles; lizards and snakes; and the tuatara.

Amphibians have moist, scaleless skin and most lay their eggs in water. The young pass through a larval stage as water-borne creatures, breathing through gills, before they become lung- or skin-breathing adults living on land. As adults they may spend time in the water too and they must return there to breed. There are three types of amphibian: legless amphibians; newts and salamanders; frogs and toads.

## Lizards

Lizards are reptiles with flattened bodies and long tails. They occur from the hottest areas to within the Arctic Circle. Most live on the ground and feed on small invertebrate prey. Generally, males have larger heads and shorter bodies than females. The male defends its territory by adopting threat postures with its head tilted upward and its throat expanded. Females lay eggs in earth or sand.

▶ **Chameleon** This tree-living lizard can change the colour of its whole body very quickly. This is ideal camouflage both when stalking prey and when hiding from predators. The pigments in the skin cells are controlled by the chameleon's nervous system. When excited, the colours brighten; when angry, the colours darken; and when frightened, the colours become paler. A chameleon can curl its tail round a branch to help it stand still. Its large eyes move independently of one another. The **common chameleon**, found in countries around the Mediterranean Sea, is 25–30 cm (10–12 inches) long. To catch insects it flicks out a long, hollow tongue with a sticky tip.

◀ **Bedriaga's rock lizard** is only found on the Mediterranean islands of Corsica and Sardinia. It lives mainly on mountains, is up to 20 cm (8 inches) long and has a pointed snout. Its young are like the adults except for a bright blue-green tail.

▶ **Frilled lizard** Also known as the frilled dragon because of its startling appearance, this slender lizard lives in forests in Australia and New Guinea. Its unusual collar of skin is 25 cm (10 inches) across. It normally lies in folds around its neck and shoulders. If threatened, the frill stands up to make the lizard look larger than it really is to its attackers. It also opens its mouth wide and hisses but it is quite harmless really. In daytime it forages in trees and on the ground for insects and other small animals. Adults are 65 cm (26 inches) long, including a tail of 45 cm (18 inches).

◀ **Gila monster** There are only two poisonous lizards – the gila monster and its close relative, the *beaded lizard* of Mexico. The venom is produced by glands in the lower jaw and is injected into prey through grooved front teeth. As the gila monster chews, the poison flows into its victim. This creature hunts mainly at night in American and Mexican deserts for mice, small reptiles, insects and birds' eggs. It moves more slowly than most lizards, although it becomes more active as the sun warms its body. A gila monster's tail shrinks as it uses up the stored fat. Its skin has shiny, bead-like scales. Gila monsters grow to 60 cm (2 feet).

▶ **Beaded lizard** This poisonous Mexican lizard has five clawed toes on each foot. It is mainly active at night, although it searches for food by day in colder seasons. The beaded lizard feeds on rodents, birds and the eggs of birds and reptiles. It has powerful jaws and when it bites it will not let go. Grooves in its lower teeth allow venom to flow into the victim. The beaded lizard's body is 80 cm (2⅔ feet) long and the tail size depends on how much fat is stored in it. This stored fat is used when no prey can be found.

▼ **Komodo dragon** This largest of all lizards is found on only a few Indonesian islands in the Pacific Ocean. Male komodo dragons grow to 3 m (10 feet) and weigh up to 91 kg (200 lb); females are smaller and lighter. This giant species evolved because it is the only meat-eater on the islands. As there are no predators, the komodo dragon does not need to run away, hide, change colour or hiss and gape. In daytime it sits motionless, watching and flicking its forked tongue in and out of its mouth. It uses its strong limbs and large claws to kill deer, wild boars and pigs. It tears meat with its jagged teeth.

▲ **Iguana** The vast majority of the 600 species of iguana live in North and South America. Most feed on insects and other small creatures. One of the exceptions is the **Fijian banded iguana** which lives on the islands of Fiji and Tonga in the Pacific Ocean. Gripping branches with its long fingers and toes, it feeds on leaves. This species is in danger of extinction due to the destruction of its forest habitat and the introduction of mongooses which eat the iguana and its eggs. The Fijian banded iguana is 90 cm (3 feet) long. Its extremely long tail is often more than twice the length of its body.

► **Gecko** This lizard lives wherever there is enough insect life to feed on and where the nights are not too cold. Most of the 675 species have huge eyes with closed, transparent eyelids. Many have pads under their toes which enable them to run up walls, walk upside down across ceilings and cling to panes of glass. Male geckos make calls of 'rok-eh' or 'gek-oh'. The **tokay gecko** of Asia is one of the largest geckos – up to 28 cm (11 inches) long. It seizes insects and mice in its powerful jaws. Females stick their eggs to upright objects and it is almost impossible to remove them without breaking the shell. To prevent dazzle, pupils in a tokay gecko's eyes narrow to four spots.

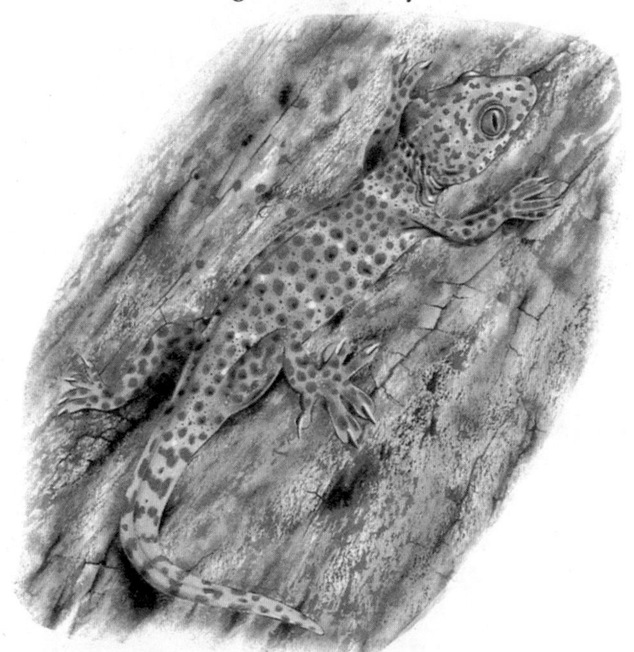

▼ **Thorny devil** One of the strangest lizards, this Australian reptile is covered in spiny scales. They are all over its body and tail, above each eye and behind its head. Each scale is enlarged and drawn to a point in the centre. Thin grooves radiate from the central peak. During cold nights, dew condenses on the scales, runs along the grooves and down into the thorny devil's mouth. These spines are an effective defence against predators. This slow-mover eats as many as 1500 ants in one meal. It grows to 16 cm (6¼ inches).

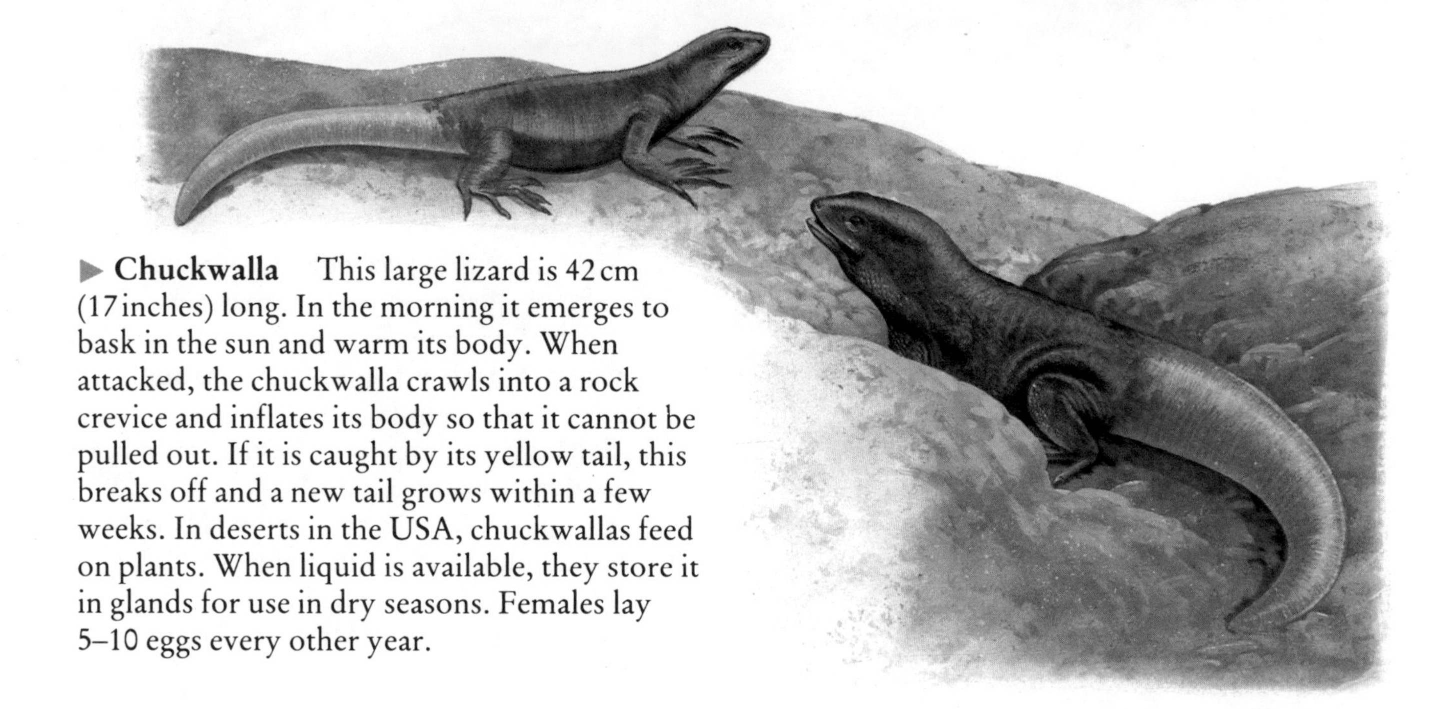

▶ **Chuckwalla** This large lizard is 42 cm (17 inches) long. In the morning it emerges to bask in the sun and warm its body. When attacked, the chuckwalla crawls into a rock crevice and inflates its body so that it cannot be pulled out. If it is caught by its yellow tail, this breaks off and a new tail grows within a few weeks. In deserts in the USA, chuckwallas feed on plants. When liquid is available, they store it in glands for use in dry seasons. Females lay 5–10 eggs every other year.

◀ **Tuatara** This reptile, which looks like a lizard, is the only living member of its group. It is similar to species which lived over 100 million years ago. Tuataras are found only in New Zealand. They have the lowest body temperature of any reptile — only 12°C/53°F. A tuatara grows slowly, but can live up to 75 years. The female lays 8–15 eggs in a hole in the soil and leaves them for about 14 months until they hatch. A male tuatara is 65 cm (26 inches) long and weighs 2 kg (4.4 lb). Tuataras live in burrows under trees and forage at night for insects, worms and snails.

▶ **Salamander** Related to the *newt*, this amphibian has well-developed limbs. The **fire salamander** is found on hills and mountains in Europe, Africa and Asia. Its bright markings give warning to enemies of its unpleasant body odours. It hunts worms, slugs and insects at night. During the day it stays beneath stones or leaves. Up to 50 eggs develop inside the female and live young are born in the water. Adults grow to 28 cm (11 inches) and live for about 24 years.

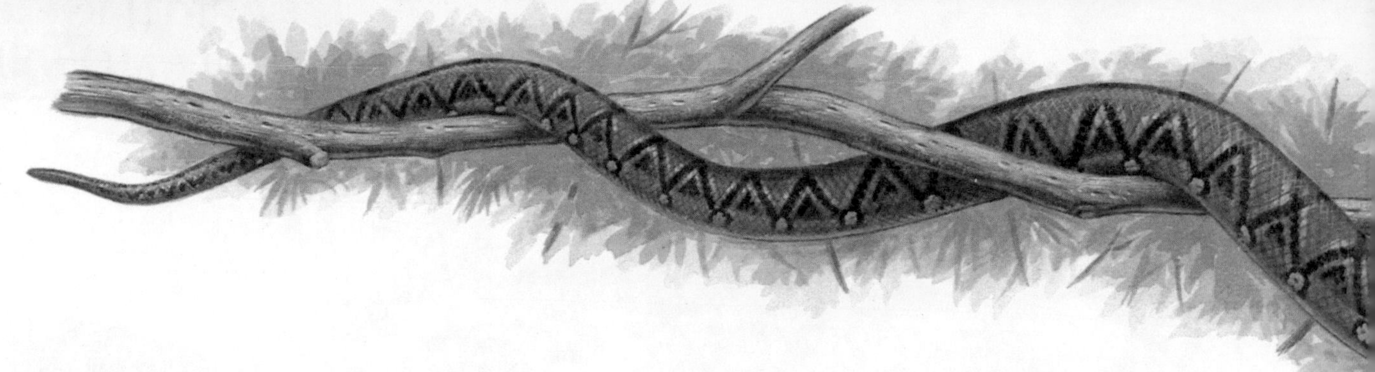

◀ **Boomslang** This very dangerous snake has three large, grooved fangs at the back of its jaws. Normally it bites lizards, particularly chameleons, as well as frogs and birds, but its venom (poison) can kill a human being. A boomslang grips its prey and chews poison into it. It releases its victim and follows its scent, waiting for it to die from the venom. This African snake smells with its tongue, which it flicks in and out. A boomslang is a tree-dweller and hunts in the daytime. Males are 2 m (6½ feet) long when fully grown.

▶ **Hognose snake** This North American species looks and acts like a poisonous snake but has no venom. When attacked, the **eastern hognose snake** flattens its head and spreads its ribs which doubles the usual width of its body. It hisses and pretends to spit poison. Then suddenly it rolls over on to its back, with its yellow underside showing. It lies still with its mouth wide open. If this 'dead snake' is turned over by a predator, it rolls back again. Hognose snakes feed on toads and insects. Females lay up to 40 eggs. Adults grow to a length of 1.2 m (4 feet).

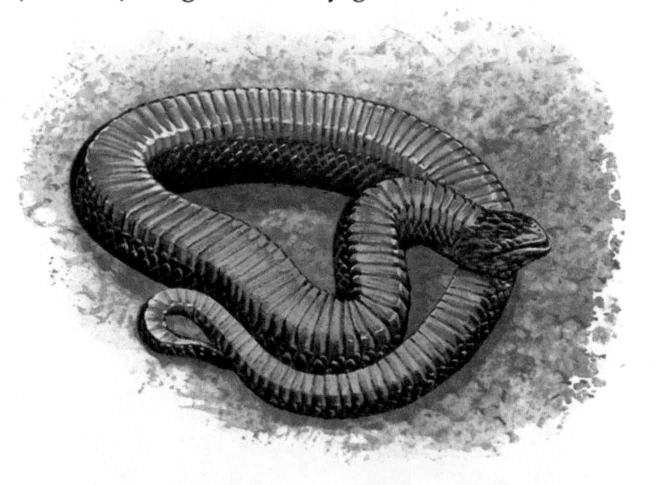

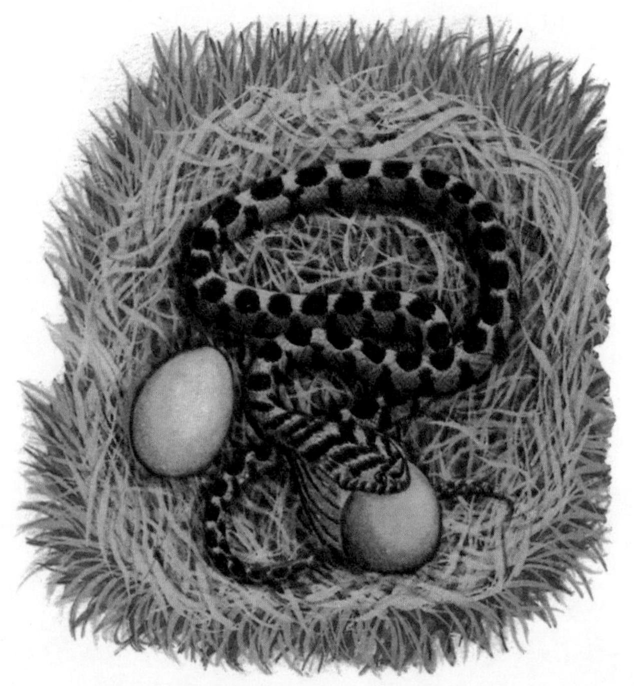

◀ **Egg-eating snake** Related to the *boomslang*, this African reptile exists entirely on birds' eggs. It searches for them on the ground and in trees, mainly at night. The egg-eating snake's mouth and jaws are very flexible and can be opened wide enough to engulf eggs twice the size of its head. It separates the two halves of its lower jaw from the top jaw. Small neck scales stand apart, exposing the skin beneath. The egg is crushed in the snake's gullet before it spits the shell out. Female egg-eating snakes have the unusual habit of laying their eggs singly, not in a clutch like other reptiles. These slender snakes are up to 75 cm (2½ feet) long.

▲ **Python** The longest snake in the world is the **reticulated python** found in South-East Asia. It grows to 9 m (30 feet) in length and weighs about 136 kg (300 lb). Pythons are not poisonous. They kill by squeezing a victim. The 20 species of python, members of the boa family, are primitive reptiles. Under their skin are tiny hip bones, the relics of legs which no longer exist. They also have two lungs, whereas more advanced reptiles have lost the left lung. Pythons are found in or near water and also spend time in trees.

◀ **Indian cobra** As well as biting, this dangerous snake can attack and defend itself by spitting two spray jets of venom 2 m (6½ feet) or more. If this poison enters an enemy's eyes, it can cause severe pain. When alarmed, an Indian cobra raises the front part of its body and spreads long neck rings and loose skin to form a hood. The markings on the back of this hood look like eyes. Indian cobras feed on rodents, lizards and frogs. The female lays up to 45 eggs in a hollow tree or termite mound. She guards the clutch until the young hatch after 50–60 days. Indian cobras grow to a length of 2.2 m (7¼ feet).

▶ **King cobra** Otherwise known as hamadryad, this is the biggest poisonous snake in the world. Adult king cobras are up to 5.5 m (18 feet) long and weigh about 7 kg (15 lb). They live near water in forests in Asia and feed on other snakes and lizards. Venom is injected into a victim through the cobra's front fangs. The king cobra's hooded head is as big as a human's. Despite its size, this agile snake flees for cover or into water when pursued. The female king cobra builds a nest of vegetation for her eggs and then lies coiled on top. This is possibly the only snake species to do this.

◀ **Boa constrictor** This non-venomous predator is found in Central and South America. It adapts to a range of climates but seems to prefer swampy forests. Adults grow to 3.6 m (12 feet) and live in tree-holes, among tree roots or in holes in the ground. Boa constrictors climb trees and are rarely found in water. They prey on birds and mammals such as rats, goats and antelopes. They encircle prey in the muscular coils of their body, suffocating or crushing the victim. Boa constictors' teeth point backward. Female boas produce live young which develop inside her body and hatch out of thin-shelled eggs as they are born.

▶ **Rattlesnakes** belong to the pit viper family. All of these nocturnal snakes have pits on each side of the head which detect heat and are used to locate warm-blooded prey. They kill by making a rapid strike, injecting venom through long, curved fangs. Rattlesnakes have a series of flattened, hollow segments on their tail. When shaken, the noise warns enemies to keep away. A new segment is added each time a skin is shed. The **eastern diamondback rattlesnake** is the most dangerous snake in the USA. It weighs up to 15.4 kg (34 lb) and is 2.4 m (8 feet) long. Its patterned skin provides camouflage.

▼ **Anaconda** Found in South American swamps, this member of the boa family is the second longest snake in the world, reaching 8.2 m (25 feet) long. It spends much of its life in muddy streams. As it can only remain submerged for about 10 minutes, the anaconda usually glides along with the top of its head above water. It waits for birds and animals to come to the edge to drink, then it wraps itself round its prey and crushes it. Up to 40 live young are born, each 66 cm (26 inches) long.

► **Grass snake** One of the most common and widespread European snakes, it swims well and lives in damp places. Females are generally longer and thicker-bodied than males. They grow to a length of 1.2 m (4 feet), although some are as long as 2 m (6½ feet). Like the *hognose snake*, a grass snake will 'play dead' if threatened. It lies on its back with its mouth open, its yellow-white belly showing and pretending to be dead. In the daytime this reptile hunts for food on land and in water. It swallows most prey alive, although a grass snake's venom will kill small animals. Frogs, toads, newts, small mammals and birds are eaten. During the breeding season, a female lays 30–40 eggs in a warm spot, such as a compost heap.

◄ **Eastern coral snake** This colourful snake – up to 1.2 m (4 feet) long – is often found near water in forests and on hills in Mexico and the USA. The red, black and yellow or white bands ringing its body may be to warn off predators. This shy reptile spends most of its time buried in sand or leaf litter. Morning and late afternoon this poisonous snake prowls in search of small lizards and snakes. When it bites prey a very powerful venom passes through fangs near the front of its upper jaw.

▼ **Water snake** About 34 snake species are specially adapted for life in water. Water snakes have small eyes and nostrils on the top of the head and when they dive, pads of skin close off the nostrils. The **plain-bellied water snake** in North America has a red belly. It moves as well on land as in water. In the morning it comes ashore to bask in the sun. This venomous reptile feeds on frogs and fish. Plain-bellied water snakes are about 1.4 m (4½ feet) long when fully grown.

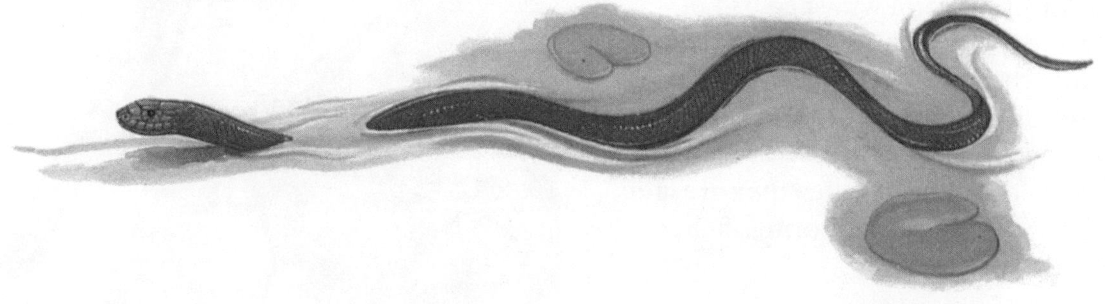

► **Strand racerunner** This lizard lives on lowland plains in South America and the West Indies. It can reach speeds of 28 km/h (17 mph) over short distances. It is always on the move, darting off in different directions. Sometimes it runs on its hind legs. Notice the extremely long toes on its hind feet and its thin, ridged tail. It uses its long, divided tongue to search for food. Strand racerunners are 30 cm (1 foot) long. They are mainly ground-dwellers but will climb low trees and bushes. The female lays four to six eggs.

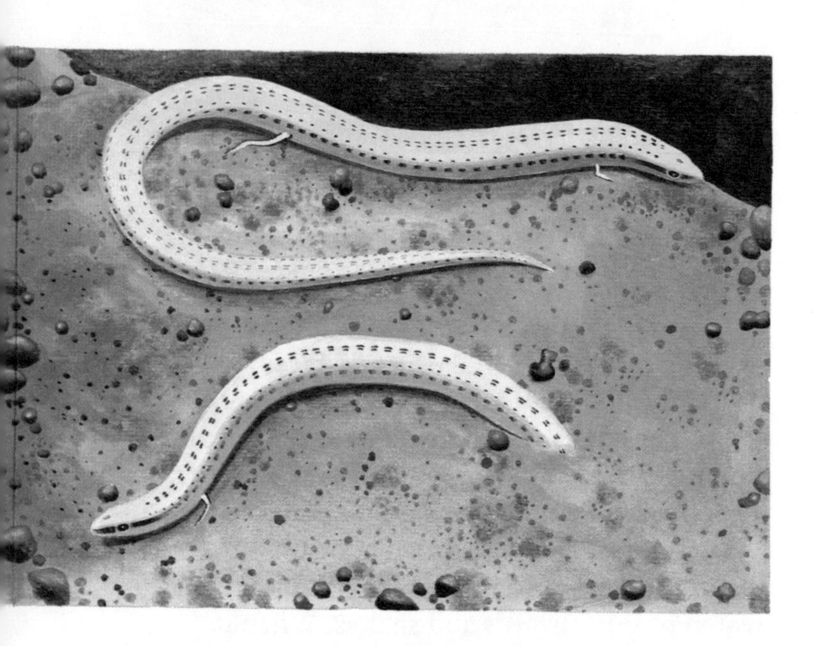

◄ **Skinks** The skink family is one of the largest lizard families, with many hundreds of species. They occur on every continent except Antarctica. Skinks have smooth scales with long, rounded bodies and tapering tails. **Florida sand skinks** have only tiny limbs which they use very rarely. This small skink – only 13 cm (5 inches) long – is an expert digger and burrower through sandhills. It uses its snout as a chisel and moves its body in a swimming motion. It feeds on termites and beetle larvae which it locates by listening to the sound vibrations they cause. Florida sand skinks mate in spring and the female lays two eggs.

► **Slow worm** This snakelike reptile gets its name because it moves more slowly than most lizards and has no visible limbs. They have traces of shoulder and hip bones showing that their ancestors had arms and legs. Slow worms are found in meadows and heaths in Europe, Asia and Africa. Females have a dark stripe on their back; some males have blue spots. They lay under rocks or logs during the heat of the day. Early morning and evening they emerge to hunt for slugs, worms and insects. When seized by an enemy, a slow worm will shed its tail. Slow worms hibernate from November through to March or April. The female is larger than the male – up to 45 cm (18 inches). She carries 8 to 20 young inside her body. They break out of their thin shells when they are laid in June or July. A young slow worm is 5–8 cm (2–3 inches) at birth.

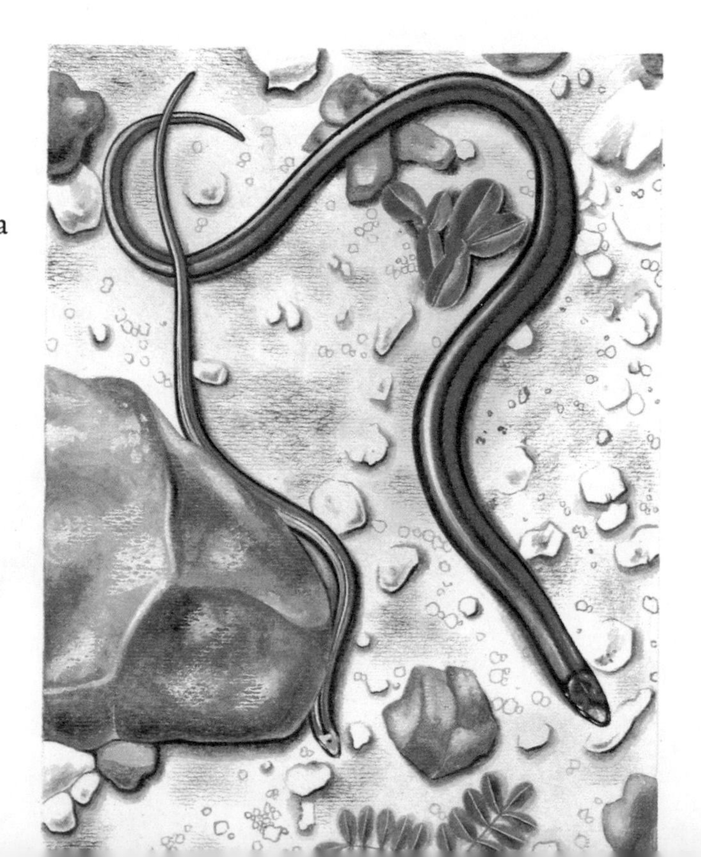

◀ **Mudpuppy** This aquatic *salamander*, related to the *olm*, lives in freshwater lakes, rivers and streams in Canada and the USA. Mudpuppies have bushy gills which vary in size. Those that live in cold water have shorter gills whereas those that live in warm, muddy water have large gills to help collect what little oxygen is available. The mudpuppy feeds mainly at night on worms, insects and small fish. The female lays up to 190 eggs. Males guard the eggs until they hatch. The larvae do not mature for four to six years. Adults measure up to 45 cm (18 inches).

▶ **Newts** The 42 species of newt and *salamander* have well-developed limbs and movable eyelids. The **rough-skinned newt** is found in or near ponds, lakes and slow-moving streams in North America. This amphibian's warty skin secretes a poison which repels most of its enemies. It looks for food both on land and in water. Rough-skinned newts have four toes on each limb. In the breeding season, the female lays her eggs one at a time on plants under water. The eggs hatch into aquatic larvae. It is not until they are adults that they develop lungs and lose their external gills.

◀ **Axolotl** The name is an ancient Mexican word meaning 'water monster'. This member of the mole salamander family lives in deep waters at high altitude in Mexico. It is now nearly extinct due to the destruction of its habitat, the introduction of predatory fish and its collection for the pet trade. The axolotl has three pairs of external gills and its dorsal fin extends down its back to the tip of its long tail and round the underside. Its legs and feet are small and weak. Female axolotls lay up to 400 eggs in water. Most of the young that hatch stay in the larval state in water; some lose their gills and change into land-dwelling amphibians. They feed on insects and small animals.
An adult grows to a length of 30 cm (1 foot).

## Tortoises

There are 39 species of this slow-moving land reptile. The strong carapace (shell) acts as protective armour. Tortoises do not usually show aggression or try to flee; they just withdraw their head and limbs inside the carapace. It is almost impossible for a predator to get at the fleshy parts of the tortoise. These herbivores are usually found in dry habitats.

Active morning and evening, they spend the remainder of the day resting. They live longer than any other vertebrate, sometimes over 150 years. Females lay white, hard-shelled eggs that are oval in shape. Young tortoises look like the adults, but have a more rounded carapace and clearer markings.

► **Hermann's tortoise** lives in lowland areas around the Mediterranean. In colder regions it hibernates under fallen leaves for the winter. The carapace is about 25 cm (10 inches) long. Females lay 2–5 eggs in soil. When the young hatch they are 4 cm (1½ inches) long.

◄ **Marginated tortoise** This reptile is found in Greece and on various islands in the Mediterranean. It looks like the *Hermann's tortoise* but the adult's carapace is flared at the edges. Marginated tortoises have coarse scales on the front of their forelimbs. Some old tortoises have dark shells and wrinkled, scarred skin.

► **Giant tortoise** This reptile can be 1.2 m (4 feet) long and may weigh over 225 kg (500 lb). Giant tortoises are found on the Galapagos islands in the Pacific Ocean. They vary in size, length of limb and shape of shell from island to island. Some have developed a shell which rises up behind the head allowing the tortoise to lift its head up and feed on the leaves of bushes. As there are no predators and no other large animals competing for grazing, these tortoises continue to survive.
They now feed on almost anything edible. Females dig a pit with their hind feet, lay about 17 eggs and then cover them with soil.

## Turtles

Turtles are superb swimmers, rarely moving far from water. They live in ponds, rivers, swamps and some species live in the sea. There are two types: those with a shell of large bones covered by bony plates; and leathery turtles whose shell consists of many small bones covered by skin. Only the females come ashore, to lay eggs.

◀ **Pondsliders** are hard-shelled turtles found in American ponds and swamps. Adults are up to 30 cm (1 foot) long. They have no teeth and feed on plants. Pondsliders bask on floating logs, often lying one on top of another. Three clutches of up to 23 eggs are laid each year. Millions of pondsliders are raised on farms and sold as pets.

▶ **Green turtles** These marine turtles move awkwardly on land, heaving themselves with both front flippers. They are found worldwide in seas where the temperature does not fall below 20°C (68°F). Green turtles feed on seaweed, sea grasses and some jellyfish. At nesting time they travel hundreds of miles back to the exact beach where they were hatched. The female sweeps away sand, digging a hole 30 cm (1 foot) deep beneath her tail. She lays an average of 100 eggs and covers them with sand. When the young hatch they have to rush down the beach to the sea before the hordes of predators, mainly birds, catch them. Those that reach the sea face more predators. Green turtles are now endangered because they are hunted and their breeding grounds have been disturbed.

◀ **Leatherback turtle** This large sea reptile, about 1.5 m (5 feet) long and with an average weight of 865 kg (1907 lb), is found worldwide usually in warm seas. Its very long flippers have a span of about 2.7 m (9 feet). The leatherback has no horny shields on its shell, no scales on its limbs and no claws. Its carapace resembles hard rubber. It feeds mainly on jellyfish. Most leatherbacks breed every other year. After laying up to 100 eggs, the female returns to the sea. Newborn turtles are 6 cm (2½ inches) long and the scales on shell and skin disappear within two months.

▲ **Alligator** This powerful carnivore has large teeth which fit into bony pits in its upper jaw. Nostrils on its broad snout tip have valves which it closes when holding prey in its open jaws beneath the surface, to stop it inhaling water. **American alligators,** 5.5 m (18 feet) long, are now protected by laws. Females scrape up vegetation with sweeping movements of body and tail to form a nest for up to 50 eggs.

◀ **Crocodiles** are the largest and most dangerous of all living reptiles. When they close their long, pointed mouths the fourth tooth on either side of the lower jaw can be seen. Alligators have this tooth hidden. Crocodiles are found in lakes and swamps. They prey on fish, birds and mammals as big as deer or cattle. Powered mainly by the tail, crocodiles move swiftly through the water. They move faster on land than is generally supposed. All crocodiles have short limbs, covered with horny skin scales. Thick bony plates on the back give added protection. Males and females look alike, but males tend to grow bigger.

◀ **Nile crocodile** The numbers of this African reptile have decreased due to the demand for skins. It preys on large mammals which come to the water's edge to drink. After seizing prey, the crocodile drowns it by holding it under water. Then it twists off flesh by spinning its own body. Nile crocodiles spend the night in water and emerge on to land early in the morning to bask in sunlight. Eggs are buried in a sandy hole and when the young crocodiles hatch, they chirp loudly and their mother uncovers them. She carries them inside her mouth to a safe nursery area. They grow to a length of 4.5 m (15 feet).

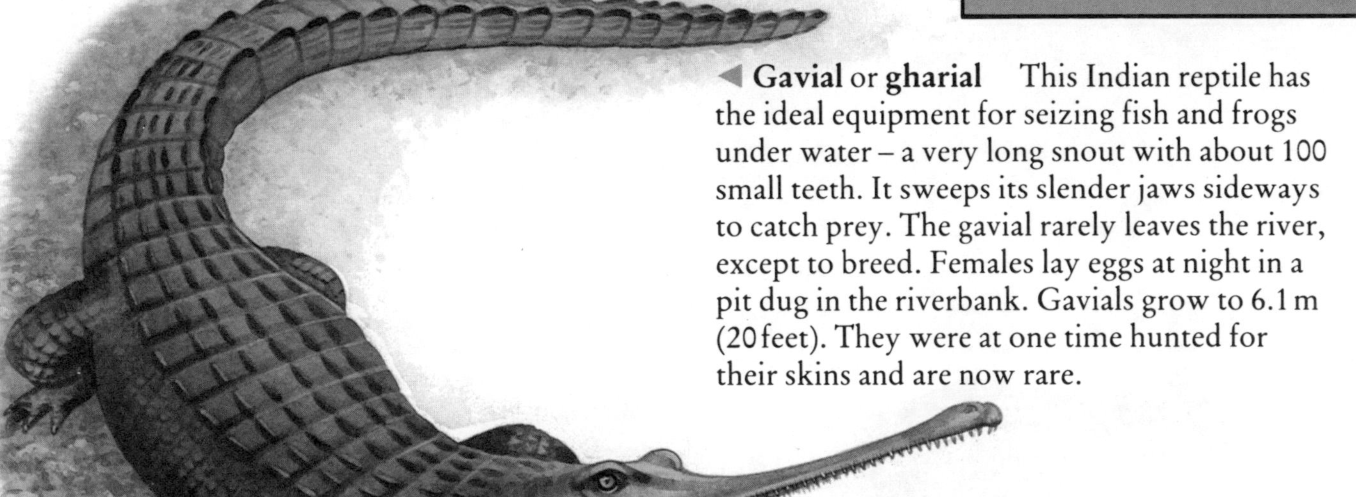

◀ **Gavial** or **gharial** This Indian reptile has the ideal equipment for seizing fish and frogs under water – a very long snout with about 100 small teeth. It sweeps its slender jaws sideways to catch prey. The gavial rarely leaves the river, except to breed. Females lay eggs at night in a pit dug in the riverbank. Gavials grow to 6.1 m (20 feet). They were at one time hunted for their skins and are now rare.

▶ **Olm** This amphibian lives in total darkness in underground streams and lakes in Yugoslavia and Italy. It is blind and so finds worms and crustaceans by smell. It has poorly developed limbs and red feathery gills behind its head. Female olms lay 12–70 eggs under a stone, although some retain a few eggs. The young are born fully developed – tiny versions of the parent but with red eyes. They grow to 30 cm (1 foot). Olms are becoming rare due to water pollution and the capture of many for the pet trade.

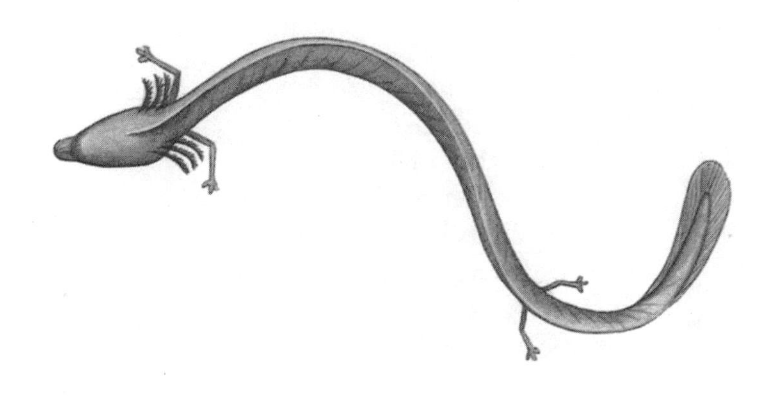

◀ **Marine iguana** This is the only lizard that lives mainly in the sea. It feeds on seaweed. Its nasal glands remove excess salt from the food and it expels this through its nose. The marine iguana cannot breathe under water but when it dives its heart beat slows, thus conserving oxygen. When not swimming, these reptiles cling to lava rocks on the Galapagos islands off the South American coast. Males butt each other when defending their own small area. They develop a green crest and red sides in breeding seasons. Females bury two or three eggs in a hole and cover them with sand. Marine iguanas grow to a length of 1.5 m (5 feet).

## Frogs and toads

Frogs and toads are amphibians. Frogs have moist, smooth skin. They use their long hind legs to move by jumping. Toads have dry, rough skin, covered in warts containing poison glands. Their legs are short and they walk rather than hop. Frogs and toads live near water and enter it to find prey or escape from danger. Adults feed on insects, spiders and crustaceans.

◀ **Common frogs** are found near ponds and swamps throughout Europe. They hibernate in mud at the bottom of ponds. Large numbers gather at the start of the breeding season. Males develop swollen pads on their forelimbs. They make 'grook grook' sounds under water to attract a female. Up to 4000 eggs are laid in large clusters that float on the surface of the water. Many eggs are destroyed or eaten. Common frogs are about 10 cm (4 inches) long.

▶ **Green tree frog** This amphibian spends most of its life in trees and bushes near ponds and lakes in Europe, Asia and Russia. Suction discs on the end of its digits help it climb. It captures flying insects with skill. The green tree frog can change its skin colour from bright green in sunlight to dark grey in the shade, providing perfect camouflage. Females lay clumps of up to 1000 eggs which float on the water. Adults are about 5 cm (2 inches) long.

◀ **Goliath frog** This is the largest living frog – 40 cm (1⅓ feet) long. It is a poor jumper and hides in deep pools when threatened. Goliath frogs can cover 3m (10 feet) in one leap, but after three or four jumps they are exhausted. Females are larger than males because of the number of eggs they carry. The heaviest one caught weighed over 3 kg (6¾ lb). Now few of these nocturnal African amphibians are left.

◀ **Midwife toads** are best known for their unusual breeding habits. Male and female mate at night on land. The female lays up to 60 eggs which the male twists in strings around his hind legs. He carries the eggs for between 18 and 49 days, depending on the temperature. At intervals, the male returns to a pool to moisten the eggs and finally deposits them in shallow water. During the day these toads hide in rock crevices, under logs or in burrows which they dig with strong forefeet. Adults of this European species grow to 5 cm (2 inches).

▶ **Common toad** This, the largest of the toads, is found in Europe, North Africa and North Asia. The average size and weight for a female common toad is 10 cm (4 inches) and 114 g (4 oz). Males are smaller and lighter – only 7 cm (3 inches) and 40 g (1½ oz). These nocturnal amphibians feed on slugs and earthworms. They hide during the day, often using the same spot. Most common toads hibernate. Large numbers gather to breed at the end of March, returning to the same pond each year. Thousands of eggs are laid in strings about 3 m (10 feet) long, coiled about water plants. In warm weather, tadpoles develop into toads in about two months. If threatened, a common toad will either hop away or will inflate its body, raise its hind quarters and sway from side to side. Common toads have been found 340 m (1115 feet) down a coalmine and 8000 m (26,256 feet) up a mountain.

## Life-cycle of the frog

Frogs and toads follow the same life-cycle.

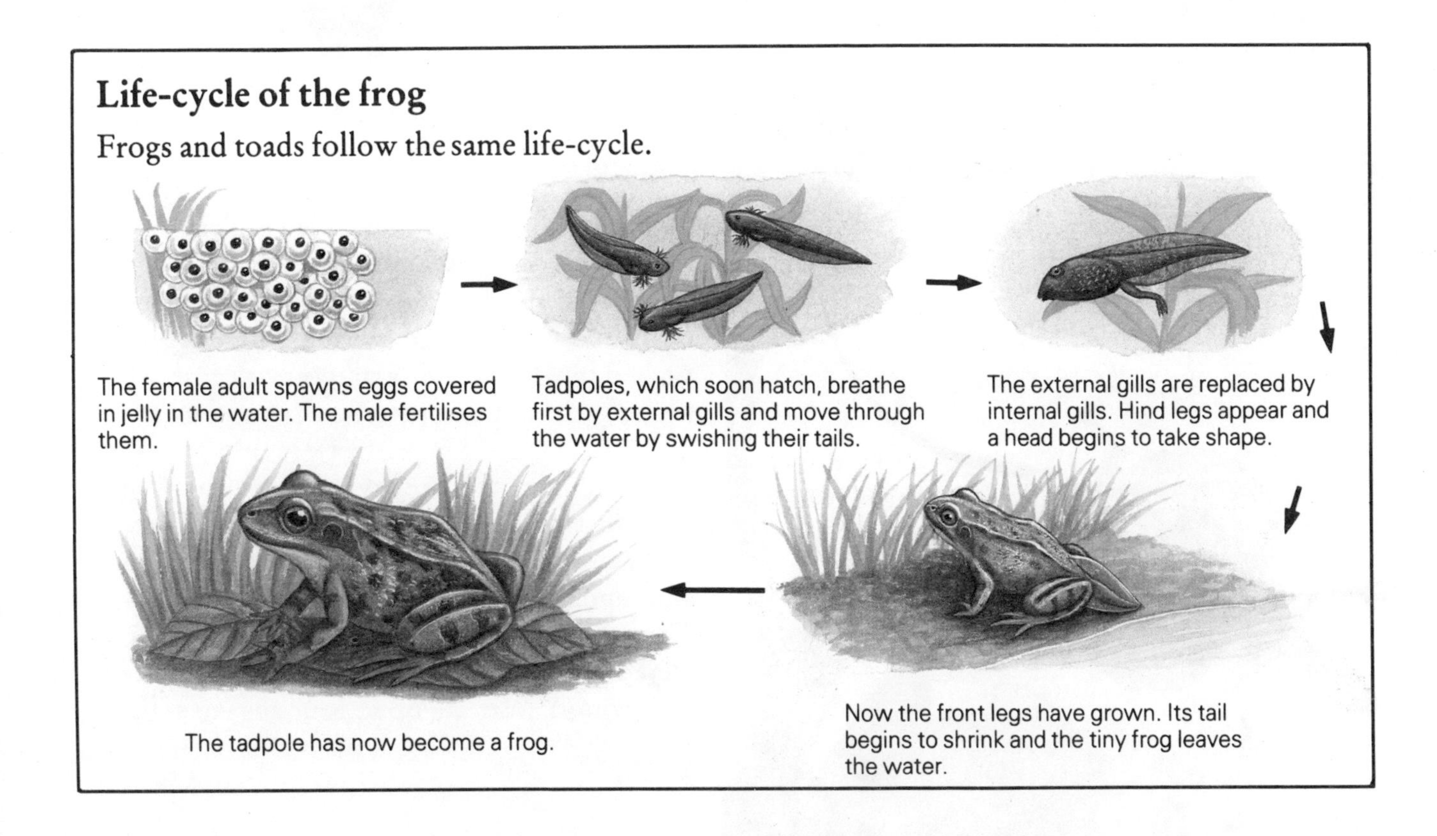

The female adult spawns eggs covered in jelly in the water. The male fertilises them.

Tadpoles, which soon hatch, breathe first by external gills and move through the water by swishing their tails.

The external gills are replaced by internal gills. Hind legs appear and a head begins to take shape.

Now the front legs have grown. Its tail begins to shrink and the tiny frog leaves the water.

The tadpole has now become a frog.

◄ **Flying snake** In Asian forests one species of snake can be seen gliding from tree to tree, from one branch to another. As the **paradise tree snake** launches itself into the air, it straightens its body. It spreads its ribs and drives all the air out of its lungs. Gliding at an angle of 50 or 60 degrees to the ground for 20 m (65 feet), it has little control over its 'flight'. The paradise tree snake has ridged scales on its belly which help it to climb vertically up tree trunks. Adults grow to about 1.2 m (4 feet) and feed on the many tree-dwelling lizards.

► **Flying frog** A species of frog that can fly is **Wallace's flying frog**. In the forests of Borneo this amphibian is adapted for gliding from tree to tree. Its long toes have wide webs between them. These flat surfaces help the frog to glide through the air. This is an effective way of escaping from predators. Sticky pads on the tips of its toes help it to hold on when landing on another branch. An adult Wallace's flying frog is 10 cm (4 inches) long. Little is known of its breeding habits. Females lay eggs in a mass of foam on a leaf or branch overhanging water.

◄ **Flying dragon** This lizard glides from tree to tree in forests in Malaysia and Indonesia. It has winglike skin flaps at the sides of its body, supported by ribs. These are tucked in when a flying dragon is at rest. Then it opens the flaps, throws itself into the air and glides for up to 18 m (60 feet). If another flying dragon enters its territory, the first one dives into the air and lands close to its rival. Then it gives an aggressive display, flicking out a flap of skin under its chin until the intruder goes. These lizards – around 20 cm (8 inches) long – feed on insects, especially ants. They breed on the ground, burying up to four eggs in soil.

# FISH

In general terms a fish is any cold-blooded vertebrate animal that lives only in the water. Dolphins, porpoises and whales live only in the water but they are warm-blooded animals and are classed as mammals not fish. Fish breathe through gills and their limbs are usually in the form of fins. Fish make up more than half of the 43,000 known species of vertebrate alive today. Most fish lay eggs. Although most eggs hatch, few survive to maturity. Sharks retain their eggs which develop inside the body and they give birth to fully formed young.

There are three very different groups of fish. There are the jawless fish which breathe through gills – these are lampreys and hagfish. Then there are two kinds of jawed fish: one boneless, which consists of sharks, rays and skates; the other bony fish and this includes all the other species from lungfish to eels to flying fish. Although not strictly fish, shellfish, octopus and squid are also included in this section.

◀ **Mudskippers** are adapted for life both in and out of water. They live in mangrove swamps in many parts of the world. The biggest is about 20 cm (8 inches) long. A mudskipper's eyes are high on its head and have a layer of clear skin to protect them. When out of water, mudskippers use pectoral (side) fins to move. They achieve rapid movement by curling their body to one side and suddenly straightening. Up to 60 cm (2 feet) can be covered with one 'skip'. Mudskippers feed and breed out of water. They eat insects and crabs. As the incoming tide brings predatory fish with it, most mudskippers retreat into mud burrows but the young climb mangrove tree roots.

▶ **Lungfish** have gills and lungs. They inhale air at the surface and absorb oxygen from water through their gills. Adults eat frogs and small fish. They shear through food with sharp plates in each jaw. **African lungfish** – up to 2 m (6½ feet) long – live in rivers and lakes. During long periods of drought they burrow into mud. After closing the burrow entrance, the lungfish curls up and covers itself with a cocoon. When the rains come again, the lungfish re-emerges. The female lays one or more eggs in a nest hole made by the male who guards them. The larvae have external gills which disappear as they mature.

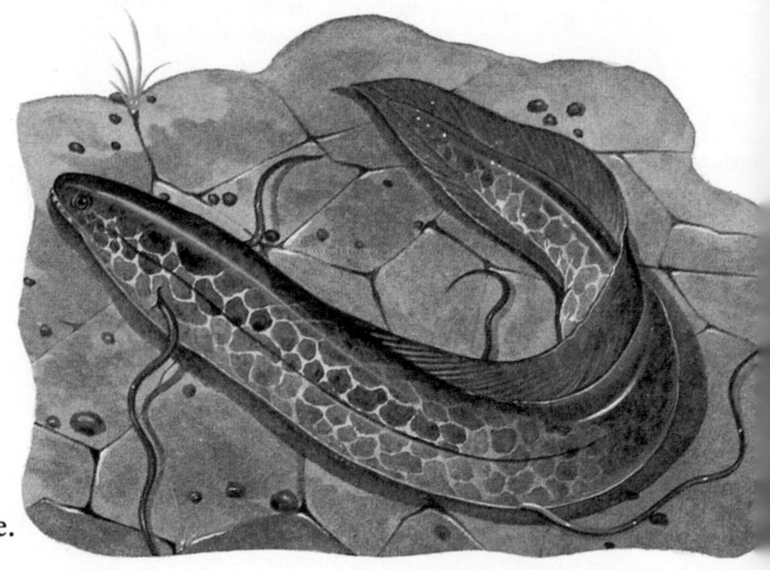

◀ **Flying fish** In the Atlantic Ocean and Mediterranean Sea lives a species of fish with four enlarged fins. The **Atlantic flying fish** swims with these fins folded against its body. It builds up speed in the water by rapidly vibrating its tail fin from side to side. As it breaks the surface, both pairs of side fins expand like wings and the fish glides for up to 90 m (300 feet). Most flights last about 10 seconds. The tropical species of flying fish has only one pair of wings. It takes shorter flights – up to 30 m (98 feet) – but reaches a speed of 56 km/h (35 mph). Adult flying fish eat other fish and grow to a length of 30–43 cm (12–17 inches).

▶ **Bichir** There are 11 species of bichir, all found in African freshwater lakes and rivers. They resemble the earliest fossil fish and have primitive features such as a swim-bladder which can be used for breathing air. They rise regularly to the surface and gulp air. Bichirs have both lungs and gills. Their long bodies are covered with hard, diamond-shaped scales. Their unusual dorsal fin looks like a row of small flags. Bichirs feed on other fish, worms and amphibians. They grow to a maximum of 40 cm (16 inches).

◀ **Sturgeon** There are 21 species of freshwater and coastal sturgeon which migrate into rivers to spawn. They have rows of bony plates along the sides of the body. Sturgeons live in seas and rivers around the European coastline. Their numbers are declining because of overfishing, pollution and obstructions in their spawning rivers. Females' unshed eggs are removed, salted and eaten as caviare. A large female sturgeon sheds between 800,000 and 2.4 million sticky black eggs in the river bed. They hatch in about a week. The young remain in the river for up to three years. Adult sturgeons feed on worms, molluscs and some fish. They weigh over 300 kg (660 lb) and are 3 m (10 feet) long.

▶ **Coelacanths** were only known as 80 million-year-old fossils until one was caught in 1938 by a trawler off the coast of South Africa. A second one was netted 14 years later near the Comoro Islands in the Indian Ocean. The internal structure of a coelacanth shows that the earliest fish had a simple heart. They have rough, thick scales. Their heavy fins are used to stir up the mud 300 m (1000 feet) down, in a search for prey. Coelacanths are ovoviviparous: this means their eggs hatch inside the mother and young coelacanths are born fully formed. Adults measure 1.9 m (6¼ feet).

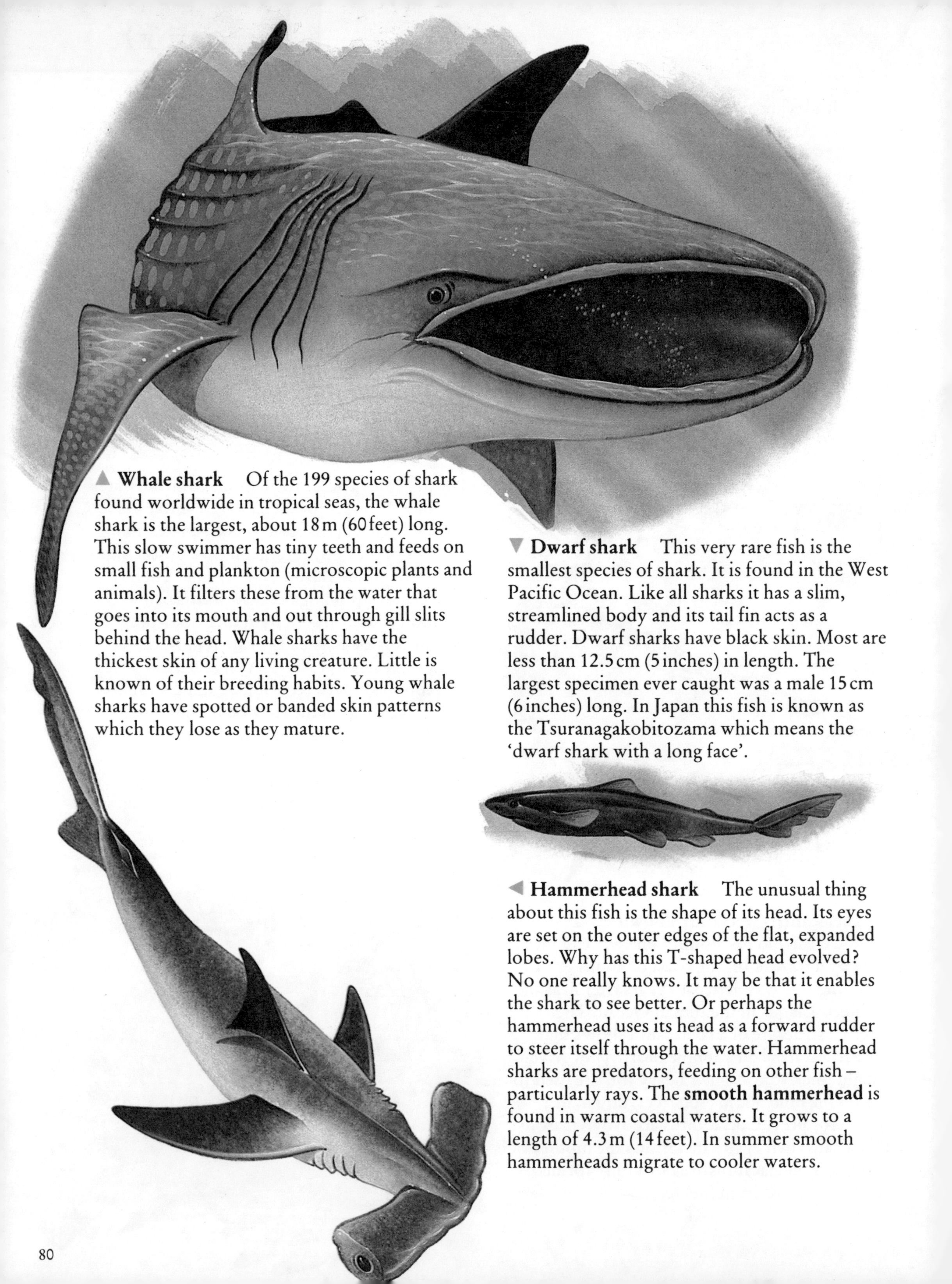

▲ **Whale shark** Of the 199 species of shark found worldwide in tropical seas, the whale shark is the largest, about 18 m (60 feet) long. This slow swimmer has tiny teeth and feeds on small fish and plankton (microscopic plants and animals). It filters these from the water that goes into its mouth and out through gill slits behind the head. Whale sharks have the thickest skin of any living creature. Little is known of their breeding habits. Young whale sharks have spotted or banded skin patterns which they lose as they mature.

▼ **Dwarf shark** This very rare fish is the smallest species of shark. It is found in the West Pacific Ocean. Like all sharks it has a slim, streamlined body and its tail fin acts as a rudder. Dwarf sharks have black skin. Most are less than 12.5 cm (5 inches) in length. The largest specimen ever caught was a male 15 cm (6 inches) long. In Japan this fish is known as the Tsuranagakobitozama which means the 'dwarf shark with a long face'.

◀ **Hammerhead shark** The unusual thing about this fish is the shape of its head. Its eyes are set on the outer edges of the flat, expanded lobes. Why has this T-shaped head evolved? No one really knows. It may be that it enables the shark to see better. Or perhaps the hammerhead uses its head as a forward rudder to steer itself through the water. Hammerhead sharks are predators, feeding on other fish – particularly rays. The **smooth hammerhead** is found in warm coastal waters. It grows to a length of 4.3 m (14 feet). In summer smooth hammerheads migrate to cooler waters.

◀ **White shark** This large carnivore is also known as blue pointer, man eater, white death and great white shark. Females are bigger than males. Adults average 4.6 m (15 feet) and are rarely over 7 m (23 feet). They feed on seals, dolphins and other sharks. As it opens its mouth, a white shark thrusts its lower jaw forward making its gape larger and more frightening. It bites off big chunks of flesh with its triangular teeth. White sharks live in open seas and migrate into coastal waters. Eggs are hatched inside the female and up to nine live young are born.

▶ **Wobbegong** This flat fish is also called the carpet shark. Most of the time it sits on the seabed near the coast of Australia. It is well camouflaged against the sandy bottom. The wobbegong uses its enlarged pectoral fins to shuffle along the seabed. It feeds mainly on fish, crushing and grinding its prey with long, thin teeth. The fringed growths round its mouth are feelers. Wobbegongs grow to 3.2 m (10½ feet) in length. The young have spotted or banded skin patterns.

▼ **Basking shark** This, the second largest fish species, gets its name from its habit of floating on the surface. It will occasionally leap into the air, possibly to remove the parasites which cling to its body. A basking shark sieves plankton from the water as it swims along with its mouth open. The largest specimen on record weighed 16 tonnes (15¾ tons) and was 12.3 m (40⅓ feet) long. Young basking sharks are 1.5 m (5 feet) long at birth.

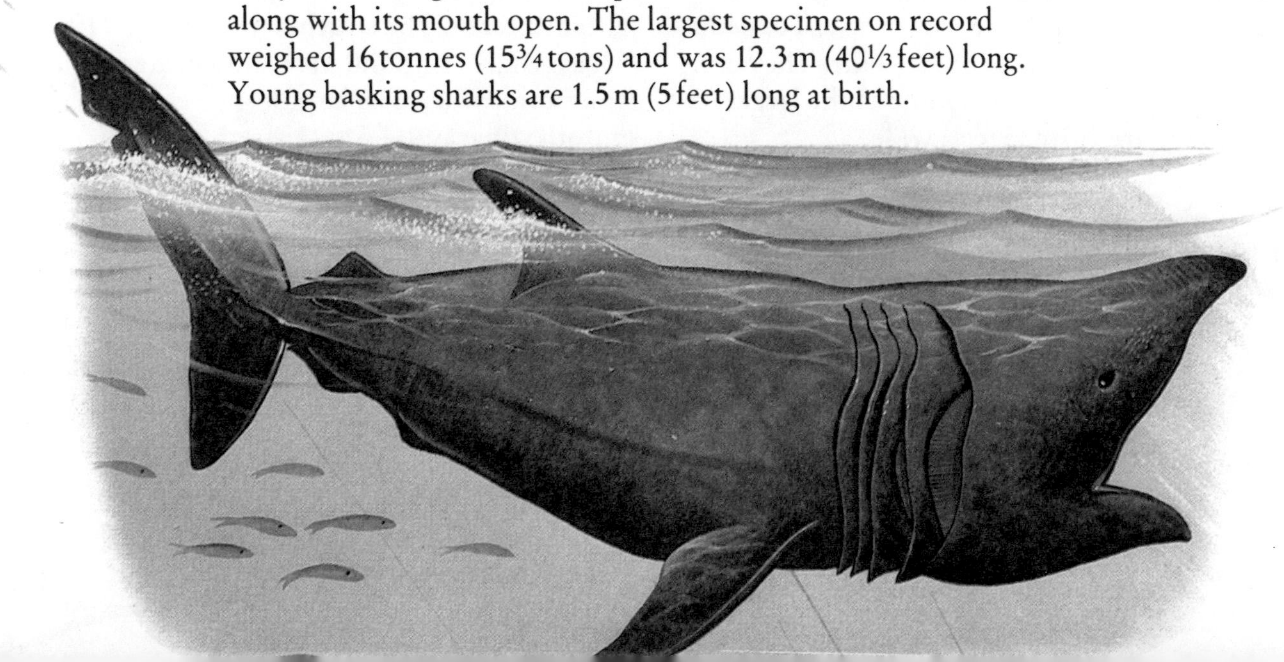

▲ **Manta ray** When swimming near the surface of the Atlantic Ocean, rays make flapping movements and appear to be flying through the water. The largest is the **Atlantic manta ray** – 5.2 m (17 feet) long and 6.7 m (22 feet) wide. 'Manta' refers to the wide fins which look like a cloak or mantle. Fleshy horns on either side of the mouth scoop in plankton. The young hatch inside the mother and are born well-developed.

▼ **Electric ray** This slow swimmer spends most of the time in shallow water, often buried in sand or mud on the seabed. It feeds on fish which it stuns with an electric shock produced by large organs at the base of its pectoral fins. Electric rays give a powerful charge of over 300 volts, enough to frighten most predators. All 30 species can produce electricity. Some live at great depth and are blind. The front part of the electric ray's body is like a round disc. **Torpedo electric rays** are 45 cm (1½ feet) wide. They are well hidden on the seabed. Some of the larger electric rays are 1.5 m (5 feet) wide.

▼ **Stingray** All species have a sharp spine near the base of the long tail. Venom is injected into a victim through this needle-like spine. A stingray wields its tail with great speed and force to drive its weapon into a victim. Its sting is painful but not fatal to humans. Stingrays usually live on the seabed, partly covered with sand. They swim away rapidly when disturbed or in pursuit of fish, crustaceans and molluscs. Food is crushed by the stingray's strong, flattened teeth. The **southern stingray** – 1.5 m (5 feet) wide – is found in shallow water around American coasts. The largest species is a smooth stingray – 4.5 m (15 feet) long.

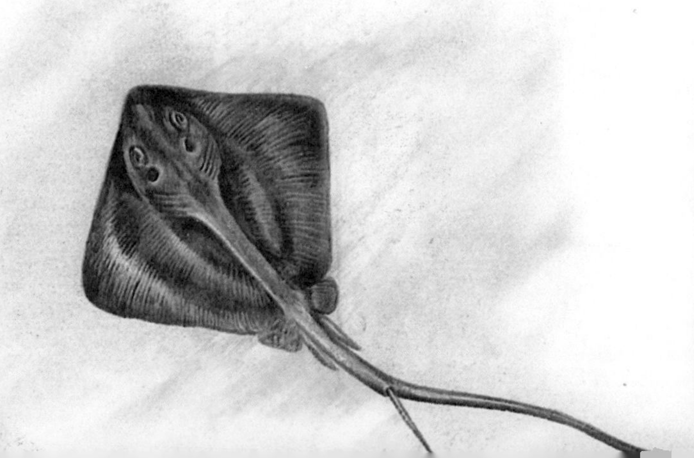

▲ **Swordfish** The only living species of its kind, this spectacular fish is found usually close to the surface in warm seas worldwide. It is around 4.5 m (15 feet) long and weighs up to 675 kg (1500 lb). Its sword is a long upper jaw. The swordfish uses this to strike out at fish and squid. Swimming at speeds of 65 km/h (40 mph) into a shoal of fish, it flails its sword from side to side. Its prey are injured or stunned and the swordfish will then eat them at its leisure.

▲ **Sailfish** Its name comes from the high dorsal fin which looks like a sail. A sailfish has a long upper jaw which is rounded, not flattened like that of the *swordfish.* It clubs victims with this weapon as it rushes through a school of mackerel. It can swim at 109 km/h (68 mph) – faster than the *cheetah* can run. Females shed several million eggs which float in surface waters in warm seas until they hatch. Their upper jaws only begin to enlarge when the young larvae are about 6 cm (2½ inches) in length. Adult sailfish are 3.6 m (12 feet) long.

▼ **Electric eel** This South American fish – 2.4 m (8 feet) long – is not a true eel but has a similar shaped body. It is the most powerful electric fish, producing a total charge of 500 volts. Much of its body is occupied by electric organs. The electric eel releases charges to kill its fish prey and for defence. It also produces pulses of electricity to help it navigate in murky waters where vision is of little use.

▲ **Gulper eel** This unusual deep-sea creature is found 1400 m (4600 feet) down in warm oceans, particularly the Atlantic. Its tiny eyes and brain are confined to a small area above the front of its huge mouth. The gaping jaws are one quarter of the total body length of 60 cm (2 feet). These jaws are joined by an elastic membrane which allows it to swallow fish. Prey are attracted by the light on the gulper eel's tail and swim towards it, straight into the gaping mouth. The larvae change into adult form at a small size, less than 4 cm (1½ inches).

▲ **Oarfish** Also called 'king of the herrings', this distinctive sea creature is found worldwide in warm oceans, 300–600 m (980–2000 feet) down. It is the largest of the bony fish, around 8 m (26 feet) long. This fish may be the source of the many stories about sea-serpents. An oarfish swims with rippling movements. Its long, silvery body has flat sides. The dorsal fin, running along the length of its body, gleams bright red in the darkness. Two pelvic fins under its chin are long and slender. They are tipped with flaps of skin which may help the oarfish in its search for small crustaceans.

## Life story of the eel

In the autumn large numbers of adult European eels (see opposite page) wriggle through the mud from stream to river and down to the sea. They set out on the 6400 km (4000 mile) journey across the Atlantic Ocean to the Sargasso Sea. This area is covered in seaweed and is rarely disturbed by storms. The adults die when they have laid their eggs. Eel larvae hatch out in the spring. In the autumn they begin to drift in warm currents, feeding on tiny creatures. They grow slowly and at 2½ years are no more than 75 mm (3 inches) long. By this time they are called elvers and are approaching the shores of Europe.

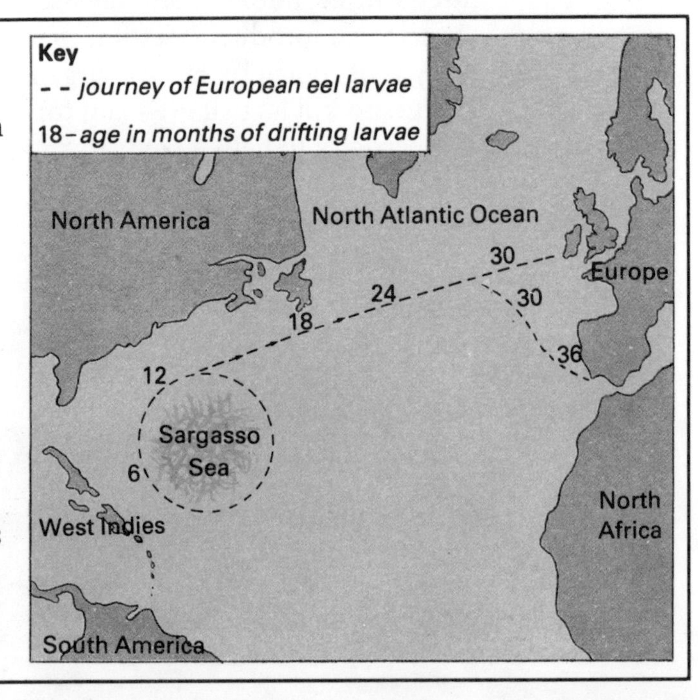

## Eels

There are more than 600 different species of eel worldwide. Although most species live in seas and oceans, 15 species are found in freshwater streams. All eels have long, slender bodies, tiny heads and long fins. Female eels produce eggs which hatch into thin, transparent larvae.

1 **European eels** live in rivers and coastal waters around North Africa and Europe. They feed mainly at night, eating almost anything – worms, frogs and crustaceans. Adult European eels are about 1 m (3¼ feet) long, weigh around 7 kg (15½ lb) and some may live for over 80 years. Many travel overland, usually on wet nights, to reach isolated ponds. The life story of this species is amazing (see opposite). When young eels approach European shores, their skin is yellowish-brown. As they mature the skin darkens, becoming almost black, and they develop a silvery belly.

2 **Moray eel** There are 100 species of this scaleless eel, widely distributed in warm oceans. An adult is 1.3 m (4¼ feet) long. It lurks in small caverns with only its head showing, ready to shoot out and grab passing fish, squid or cuttlefish. Moray eels deliver savage bites with their sharp teeth.

3 **Conger eel** In shallow waters in the North Atlantic and Mediterranean, this large eel feeds on crabs, fish and cuttlefish. It is nocturnal, hiding by day under rocks. Females are larger than males and grow up to 3 m (9¾ feet). During its breeding season, a conger eel's stomach shrinks and its teeth fall out. After spawning millions of eggs, the adults die. The transparent larvae drift for one or two years, developing into adults.

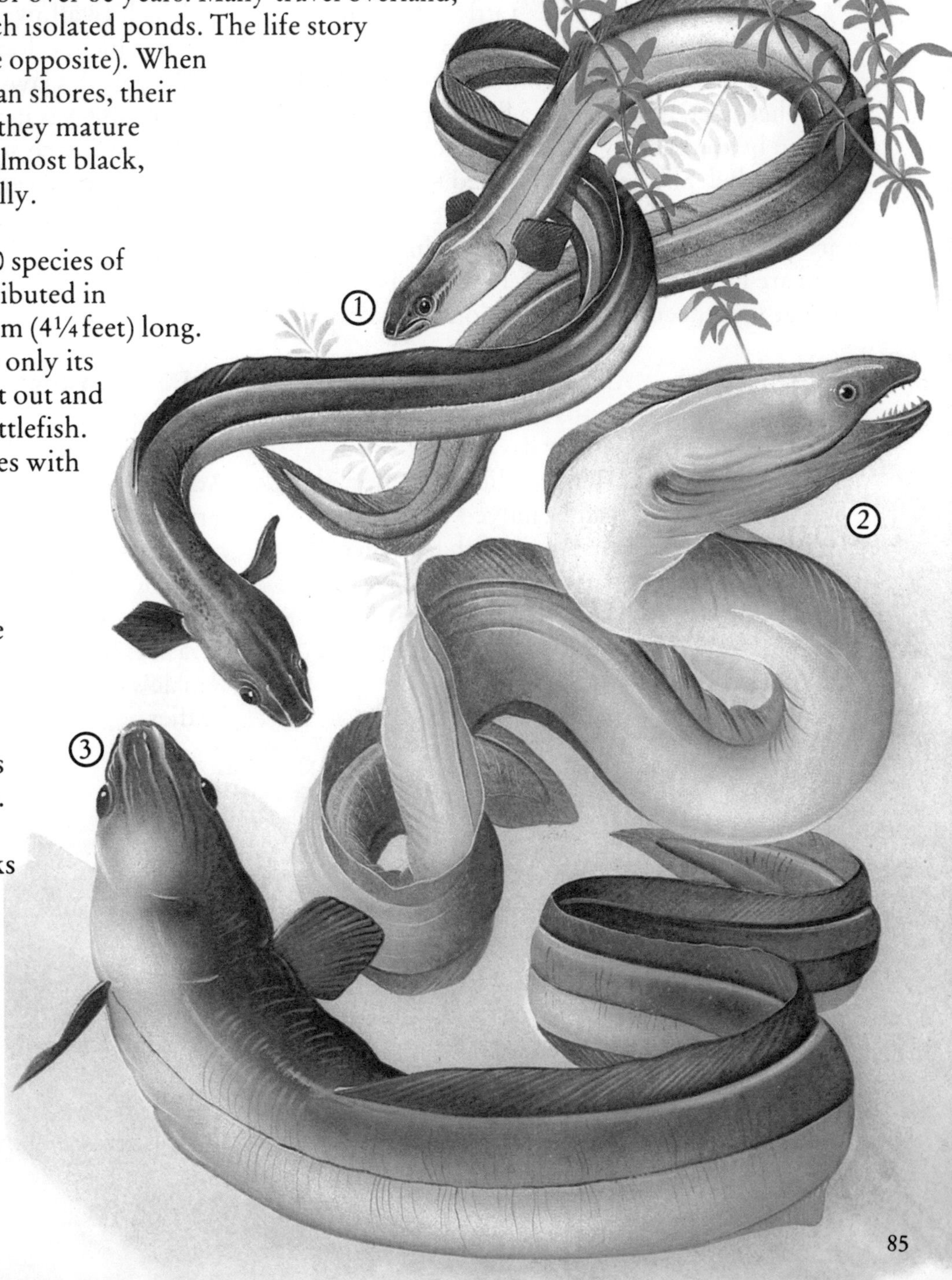

▲ **Deep-sea dragonfish** Little is known about fish that live in the dark depths of oceans. The female deep-sea dragonfish lures small fish into her open mouth with a light-producing organ below each eye and a luminous (shining) chin barbel. She has thin, hooked teeth and grows to over 30 cm (1 foot) in length. Males are brown, have neither teeth nor chin barbel and are never longer than 5 cm (2 inches). Their larvae look peculiar, with eyes on stalks.

▲ **Black swallower** This small fish – only 15 cm (6 inches) long – lives at depths of 2440 m (8000 feet). Its jaws, full of sharp teeth, can become huge as it engulfs prey up to twice its own size. Its stomach stretches like a bag. A large meal lasts for several days. Like over 1000 other fish species, the black swallower has light organs on its body.

▼ **Angler fish** All 215 species have large heads with wide mouths filled with rows of sharp teeth. The majority have a light organ on the top of the head. This modified fin, tipped with a flap of skin, can be moved to a position in front of the mouth. Angler fish feed on other fish, squid and crustaceans. Females have teeth which shine in the dark. In one species the female (shown below) is 15 cm (6 inches) long, 20 times the length of the male.

▼ **Deep-sea angler fish** The female of one of the 20 species of deep-sea anglers has a branched chin barbel that looks like seaweed. Adult females are 7 cm (3 inches) long. Male larvae of both the species of angler fish shown here grow slowly. When mature, they look for a female of the same species and attach themselves to her by biting her skin. They are parasites feeding on nourishment in the female's blood. If a male fails to find a mate he dies.

▲ **Piranha fish** These predators are the most ferocious freshwater fish in the world. They live in large rivers in South America. Piranhas use their razor-sharp teeth to bite chunks out of other animals. Their jaw muscles are so powerful that they can remove a human finger or toe. The 'feeding frenzy' associated with these fish only occurs when groups of 20 or more smell a victim's blood. **Red piranhas** grow to around 30 cm (1 foot) in length.

▼ **Porcupine fish** are covered with long, sharp spines which normally lie flat. When in danger the porcupine fish takes in air, blowing up like a balloon. The spines stand up and stick out. It is almost impossible for predators to tackle this moving pin cushion. This compensates for the fish's lack of speed and mobility. Two fused teeth in each jaw crush hard-shelled prey such as sea urchins and crabs. **Spotted porcupine fish** are 90 cm (3 feet) long.

▲ **Archer fish** Living in freshwater streams in Indonesia, the archer fish uses a clever technique for catching insects. When it spots an insect, it presses its tongue against a groove in the roof of the mouth, forming a narrow tube. It jerks its gill covers shut and shoots out droplets of water. Adults – 25 cm (10 inches) long – can knock down an insect 1.5 m (5 feet) away. If the prey is close to the surface, the archer fish jumps out and snatches it.

▼ **Clownfish** swim in and out of the tentacles of sea anemones on coral reefs in the Pacific. They lay eggs at the base of the anemone. These are then safe from enemies. When other fish try to attack the clownfish, the sea anemone stings with its tentacles. Both anemone and clownfish eat the victim. The clownfish is protected from the anemone by its own body mucus which lacks the protein that normally triggers off the stinging cells. **Orange clownfish** are 8 cm (3½ inches) long.

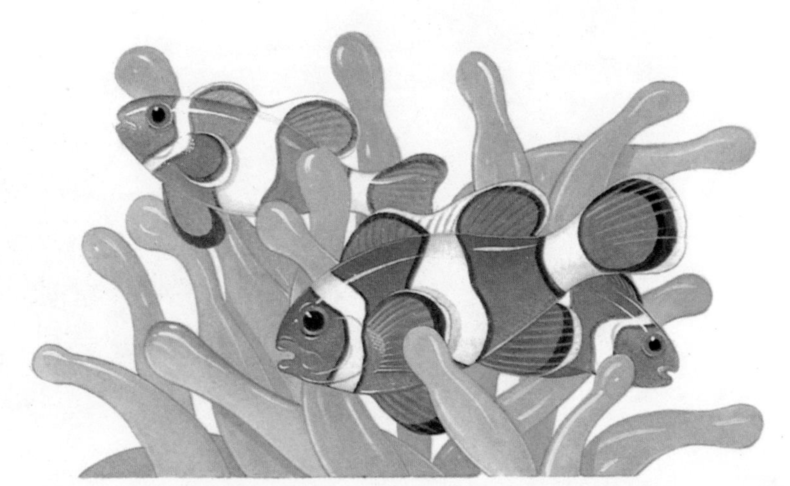

► **Ocean sunfish** This fish is unlike any other. Its body is circular when seen from the side and instead of a tail it has a frill at the rear. Under its scaleless skin is a thick layer of tough gristle. Two fused teeth in each jaw make a strong beak. The ocean sunfish feeds on tiny jellyfish and crustaceans. Its young have long spines on their bodies. These gradually disappear as the adult shape develops. Adults are 3 m (10 feet) long and 4 m (13 feet) across. The ocean sunfish was given its name after it was spotted supposedly basking in the sun in surface waters. It is now thought that these were only sick ocean sunfish.

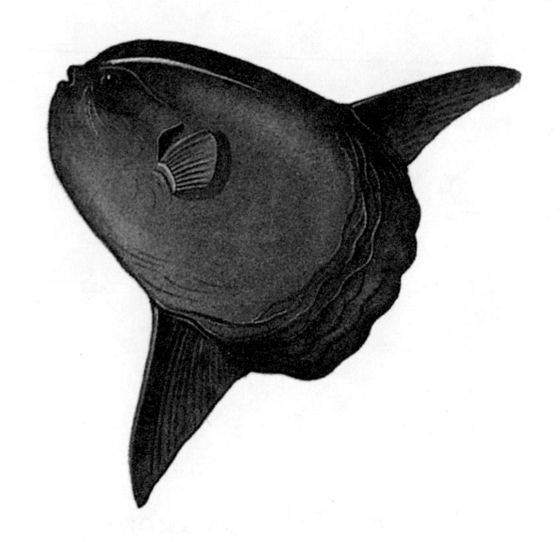

◄ **Hatchet fish** By day hatchet fish are found between 400 m (1300 feet) and 600 m (2000 feet) down in tropical seas. They migrate nearer to the surface each night in search of food. Rows of light-producing organs on their sharp-edged belly are arranged in a different pattern on each species. This enables them to recognise their own kind from below. Hatchet fish have large, glowing eyes. From the front they look frightening. From the side the silvery body is shaped like a hatchet. A hatchet fish is no more than 10 cm (4 inches) long.

▼ **Viper fish** This long, thin predator feeds on smaller fish which it follows when they make their nightly trips to the surface. **Sloane's viper fish** is one of six species of deep-sea viper fish, all with fang-like teeth. The skull is adapted to increase the gape of the mouth. An adult Sloane's viper fish is 30 cm (1 foot) long. The first ray of its dorsal fin is extra long and has a light organ in its tip.

◀ **Remoras** These fish attach themselves to other fish and turtles by a pad on their head. The suction is so strong that it takes great force to lift them off. They feed on scraps of food that the fish or turtle spills. Remoras have a large mouth with sharp teeth and a tongue. The **sharksucker remora** is the best known species. It is about 1 m (3¼ feet) long. Compare this with the length of the blue shark on which it is riding – 3.8 m (12½ feet). Some islanders in the Pacific occasionally use remoras to catch turtles. They fit a line to the fish's tail, lower it into the sea and wait for it to attach itself to a turtle.

▶ **Angelfish** are brilliantly coloured fish found in warm seas. Angelfish change colour and pattern as they grow. They feed by scraping small organisms off corals and rocks. The **queen angelfish** has a deep, slim body. Like *Moorish idolfish*, its dorsal and anal fins extend past the tail fin. This species has a row of small spines near each gill cover. An adult queen angelfish is up to 45 cm (1½ feet) long.

◀ **Moorish idolfish** are brightly coloured tropical fish found in shallow waters in the Indian and Pacific Oceans. Their bodies are compressed which means they look thin when viewed from the front. The dorsal and anal fins are pointed and swept back. A Moorish idolfish is 18 cm (7 inches) long and 23 cm (9 inches) deep. The young have a sharp spine at each corner of the mouth. These drop off as they mature. They feed on coral.

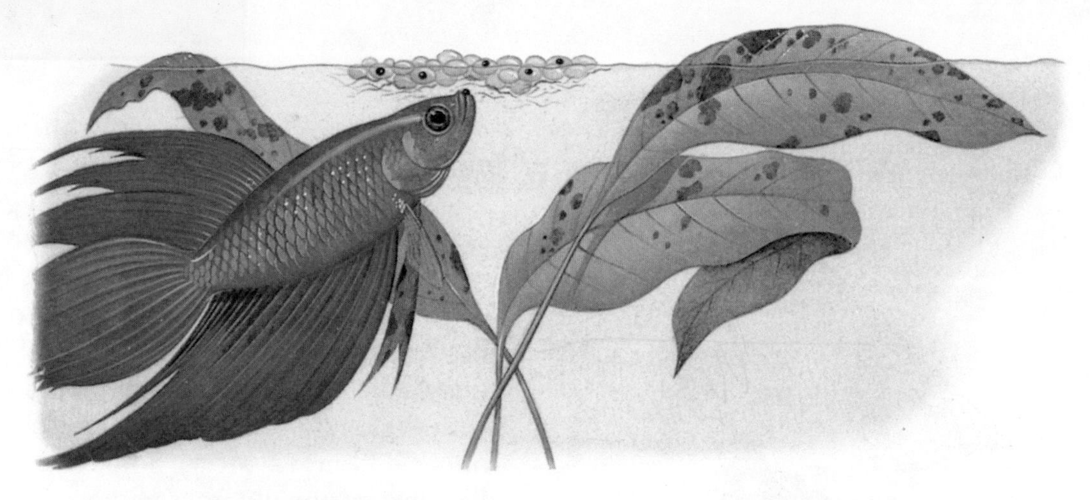

▲ **Siamese fighting fish** In clear, weedy ditches and ponds in Malaysia, rival males fight over territory. Then each male builds a nest of air bubbles which stick together and float on the surface. Next, the male displays to a drab, yellowish-brown female. If she fails to respond, he bumps and bites her until she does. As the female sheds her eggs, the male catches them in his mouth. He sticks them to the bubble nest and guards them until the young hatch out. Siamese fighting fish grow to a length of 6 cm (2½ inches).

▼ **Goby fish** are solitary, living near the seabed. They lay eggs in clusters attached to shells or stones and care for the eggs. One species of goby fish – 8 cm (3½ inches) long – shares a burrow with a pistal prawn. The blind prawn digs out the burrow, shovelling with its pincers. Then the goby acts as sentry at the opening, watching for sea snakes and other enemies. The goby taps its tail to signal to the prawn when it is safe to leave.

▼ **Sea catfish** live in Atlantic coastal waters and estuaries. Large shoals swim together at night, feeding on crabs, shrimps and fish. As the female lays her eggs, the male takes them into his mouth. He carries them carefully for about a month until the young hatch out. During this period he is unable to eat. The young fish continue to use the male's mouth as a refuge until they are about 5 cm (2 inches) long. Adult sea catfish are 30 cm (1 foot) in length.

▲ **Nile mouthbrooder** This sturdy fish lives in rivers and dammed-up pools in Africa. It has a small mouth and tiny teeth. Nile mouthbrooders feed mainly on plankton, although insects and crustaceans are also eaten. Adults are around 50 cm (20 inches) long. The female of the species carries her developing eggs inside her mouth. This keeps them safe from other fish. Even when the young have hatched, they will return to their mother's mouth when danger threatens.

▼ **Sunfish** live in freshwater rivers and streams. They scoop out a shallow, round nest. The male carefully removes pebbles, leaving the sand or gravel. The female lays up to 1000 eggs in the nest and swims away. Then the male guards the eggs, fanning fresh water over them with his tail. When the young hatch, they are led away by the male. The **bluegill sunfish** of North America is closely related to the *perch*. It is 20 cm (8 inches) long.

▼ **Pinecone fish** This small fish is also known as pineapple fish and knight fish. Its plump body is encased in heavy scales which overlap one another. Stout spines on its back lean to left and right. Pinecone fish move in schools near the bottom of the Indian and Pacific Oceans. They are often washed ashore after storms. Under the lower jaw two light-producing organs are full of luminous bacteria. Adult pinecone fish are 12.5 cm (5 inches) long.

## Coral Reefs

Coral reefs look like underwater gardens full of brightly coloured plants. In fact corals are carnivorous animals called polyps. They are similar to sea anemones but with skeletons. Most corals are not solitary but grow in colonies. As new polyps grow, others die forming a coral reef with a base of millions of dead polyp skeletons and a layer of living polyps covering its surface. Coral reefs are found in clear, shallow water in tropical areas. The largest and best known is the Great Barrier Reef, off the coast of Australia. It is 2027 km (1260 miles) long and was laid down thousands of years ago. Many different species of plants and animals are attracted to coral reefs, including several kinds of fish.

1 **Lionfish** have long, spiny fins along their back and large side fins that look like fans. Their striped bodies warn enemies that they have venom in their grooved spines. The lionfish's poison can have a serious effect, even on humans. It is a good form of defence against other predators. An adult lionfish is 37 cm (15 inches) long and is found near coral reefs in the Indian and Pacific Oceans.

2 **Cleaner wrasse** clean food and parasites from the teeth and bodies of larger fish. Here the tiny cleaner wrasse is cleaning a fish called sweetlips. The shade of blue varies according to the wrasse's age and mood. It is 10 cm (4 inches) long and usually sleeps buried in sand. It can change sex, from male to female and vice versa. It is known as a sea swallow in South Africa and a blue streak in Australia.

3 **Sea urchins** are found on coral reefs around the world. There are 800 species of this invertebrate. They move slowly over boulders, gnawing at algae and barnacles with teeth that project from a mouth in the centre of their undersides. Their spines are movable and produce a painful wound. Round sea urchins, found off the Californian coast of America, are about 18 cm (7 inches) in diameter. The largest species, found off Japan, is 30 cm (1 foot) wide.

4 **Sea anemones** look like harmless plants, swaying gently in the water. In fact, this carnivore's tentacles have stinging cells that can paralyse small fish. Sea anemones rely on prey blundering into them for they are anchored to the sea floor. The largest species, found on the Great Barrier Reef, is 60 cm (2 feet) in diameter. Sea anemones have more than eight tentacles and reproduce by budding.

5 **Parrotfish** live above steep ridges of coral reefs. They move up and down on the tide. Their teeth are fused into a sharp beak, like that of a parrot. Round teeth at the back of the mouth grind the algae on which it feeds. Males and females of the 80 species have different colour patterns. The **rainbow parrotfish** is 1.2 m (4 feet) long. It sometimes secretes a cocoon of mucus around its body at night to protect it.

6 **Spiny lobster,** also called crawfish or rock lobster, usually lives in holes on the sea bottom. This carnivore forages for prey at night. After several weeks in one place, a spiny lobster moves to a new location. It is about 50 cm (1⅔ feet) long and weighs up to 22 kg (48 lb). A spiny lobster may live for over 100 years. Its flailing antennae can inflict painful injuries. A hard outer case and spines protect it from attack.

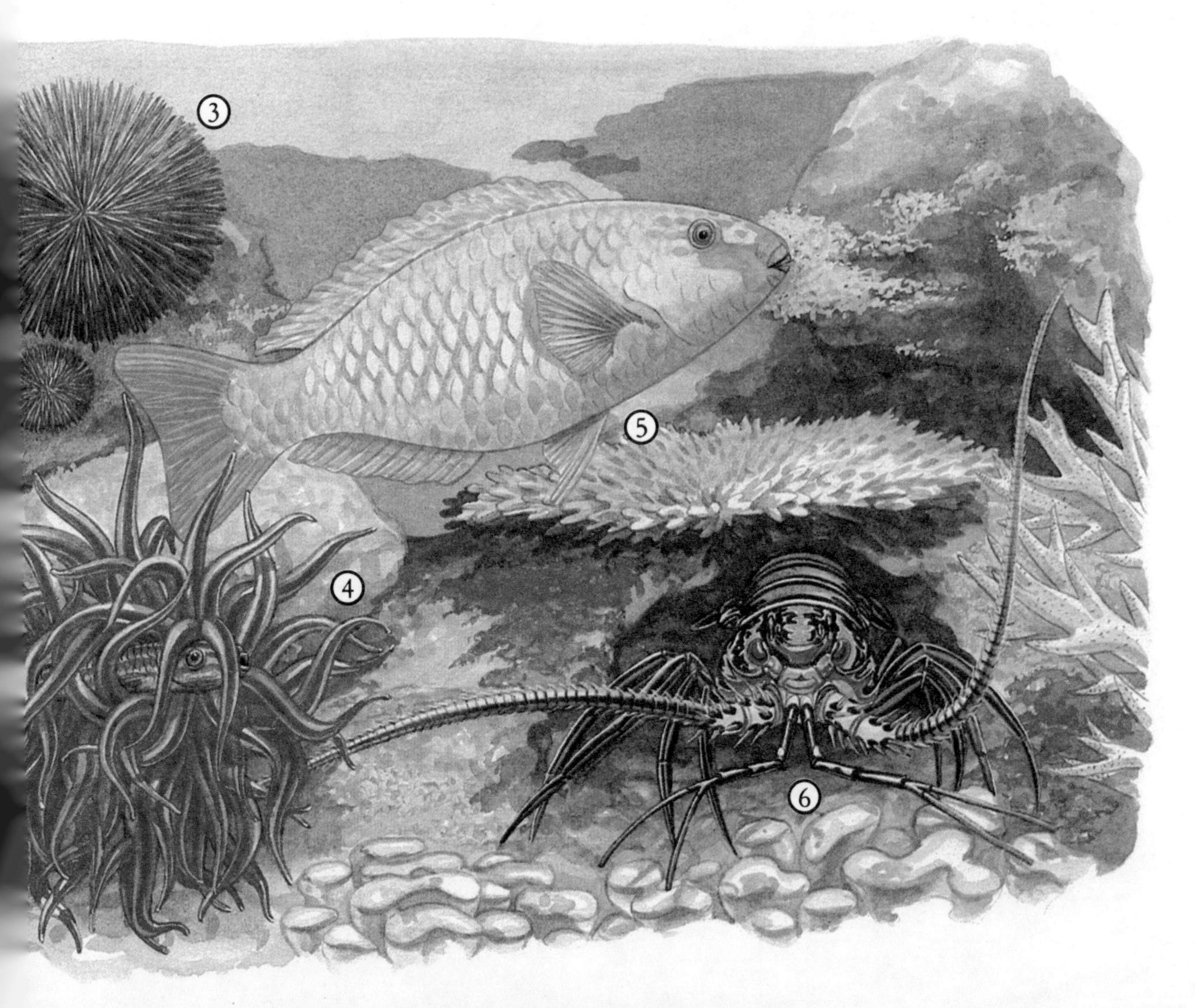

▲ **Cod** The 648 species of codfish include *haddock* and *ling*. Their bodies are covered with small scales. The cod has a large mouth full of small teeth. It feeds on many kinds of fish, worms and crustaceans. The barbel on its chin has taste buds in it. Large schools of cod live in the North Atlantic Ocean. A female releases over six million eggs into the water. Only one in a million reaches adulthood – 1.2 m (4 feet) long. Very many never hatch, others are eaten by various sea creatures and 400 million fish are caught in nets each year.

► **Haddock** live close to the seabed in coastal waters in the North Atlantic Ocean, to a depth of 300 m (1000 feet). They look like a small version of cod, especially with their three dorsal fins and two anal fins. They have a distinct black mark on each side. The breeding season is between January and June. The eggs are left to float in surface waters. Young haddock often shelter among the tentacles of large jellyfish. They grow to a length of 75 cm (2½ feet).

▼ **Ling** have long, slim bodies with two dorsal fins and one barbel on their chins. Ling are common fish around rocks and wrecks as deep as 400 m (1350 feet) in the Atlantic Ocean around Iceland, Norway and the Bay of Biscay. They feed mainly on fish and crustaceans. Ling grow to a length of 2 m (6½ feet). They breed from March to July and the female sheds many millions of eggs. These float in surface waters until they hatch into larvae.

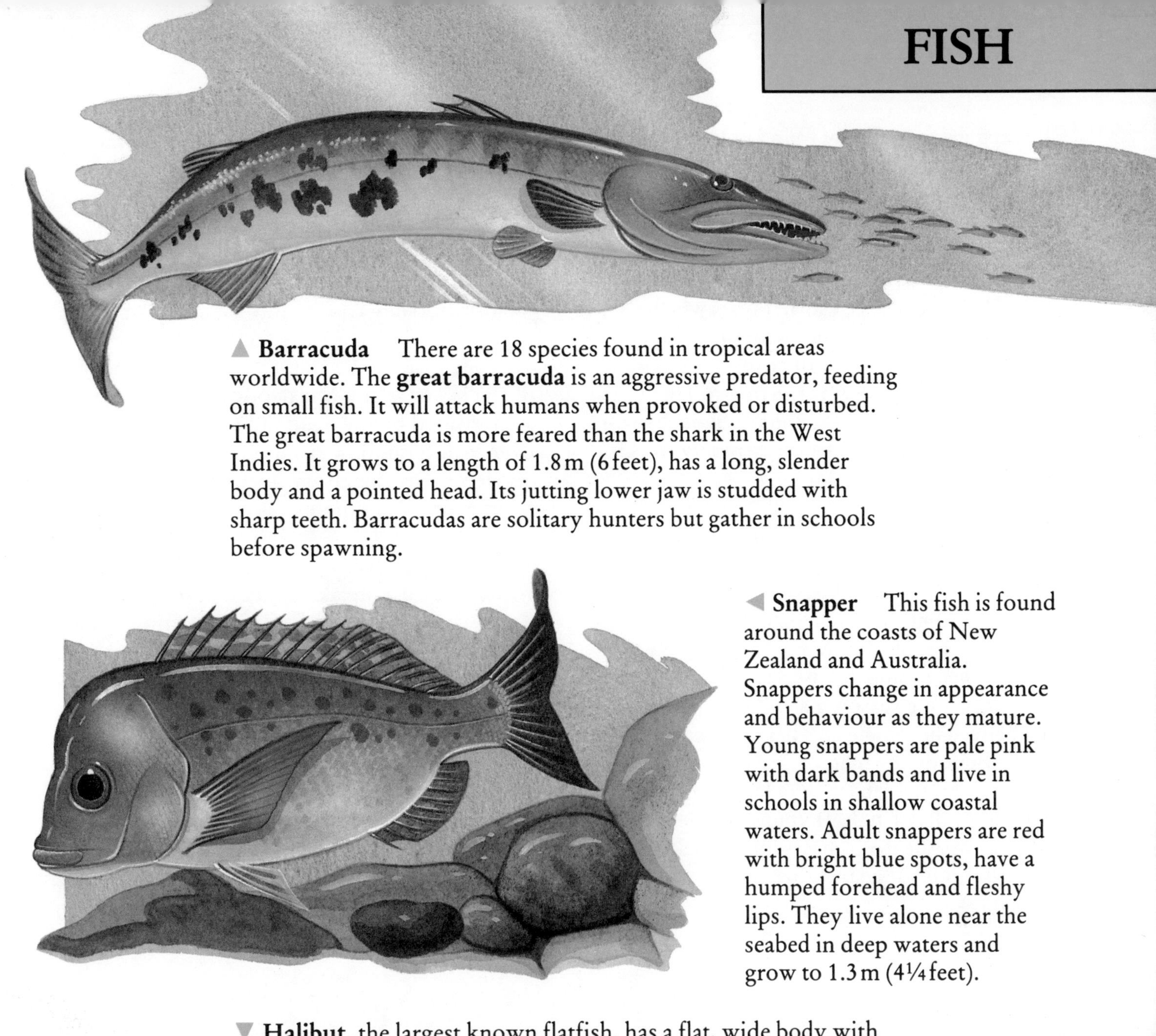

▲ **Barracuda** There are 18 species found in tropical areas worldwide. The **great barracuda** is an aggressive predator, feeding on small fish. It will attack humans when provoked or disturbed. The great barracuda is more feared than the shark in the West Indies. It grows to a length of 1.8 m (6 feet), has a long, slender body and a pointed head. Its jutting lower jaw is studded with sharp teeth. Barracudas are solitary hunters but gather in schools before spawning.

◄ **Snapper** This fish is found around the coasts of New Zealand and Australia. Snappers change in appearance and behaviour as they mature. Young snappers are pale pink with dark bands and live in schools in shallow coastal waters. Adult snappers are red with bright blue spots, have a humped forehead and fleshy lips. They live alone near the seabed in deep waters and grow to 1.3 m (4¼ feet).

▼ **Halibut**, the largest known flatfish, has a flat, wide body with both eyes on the right side. Females are larger and live longer than males. In 1957 a female halibut caught in the North Sea was 3 m (10 feet) long, weighed 229 kg (504 lb) and was 60 years of age. Females lay two million eggs at a time. These drift near the surface until they hatch. The young do not reach maturity for 10–14 years. **Atlantic halibuts** live on the bottom of the ocean, 1500 m (5000 feet) down. They are fierce predators and can camouflage themselves by changing colour.

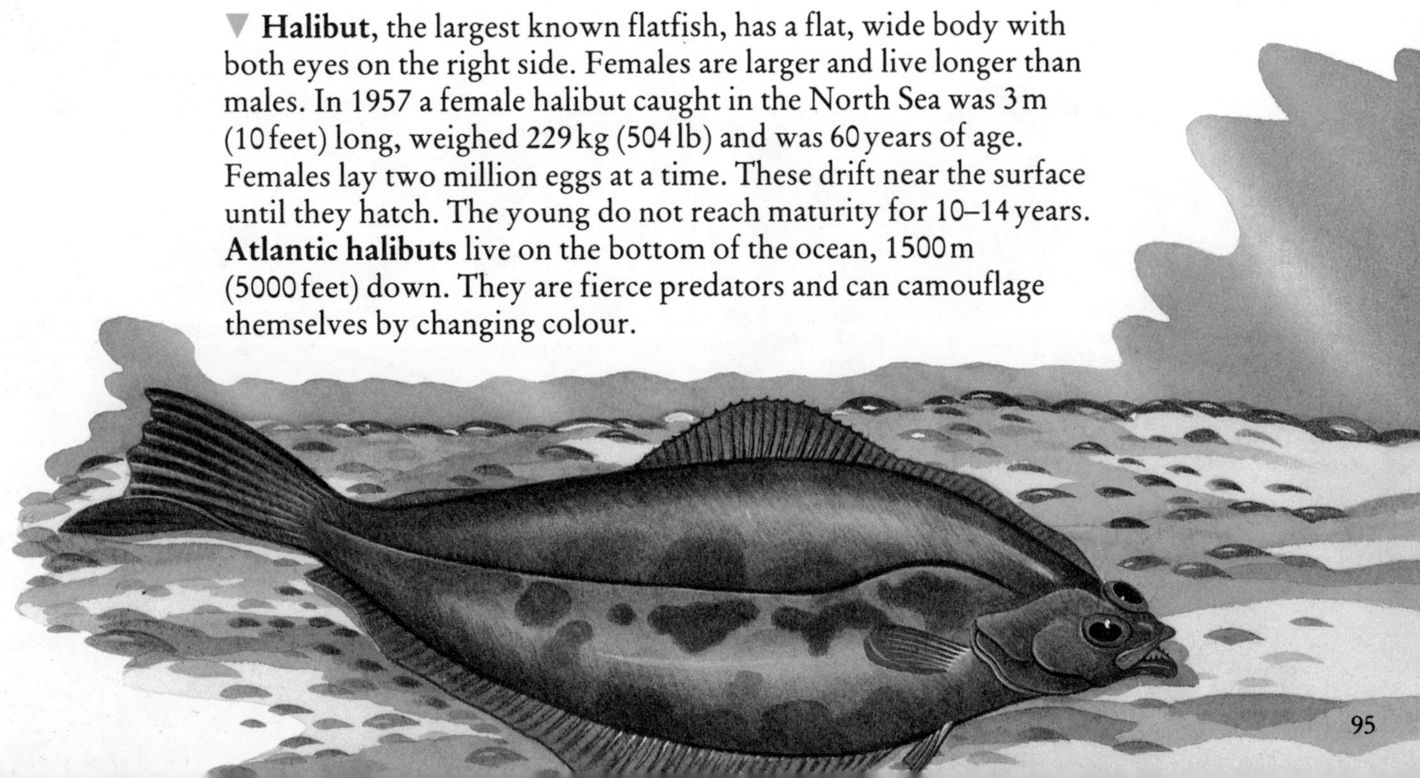

1 **Carp** were originally found in the Black and Caspian Seas, but they have been introduced into lowland lakes and rivers in many countries. They prefer densely weeded, warm water. Carp feed on insect larvae, molluscs, crustaceans and vegetation. Their hearing is acute due to a series of small bones which connect the swim-bladder with the inner ear. **Mirror carp** have a row of large scales along their sides. Some adult carp grow to 1 m (3¼ feet) and weigh 30 kg (66 lb). Some live for 50 years, although the tooth carp only lives for eight months in the wild.

2 **Sticklebacks** live in lakes, rivers and coastal waters around the world. They have spiny fins. The **three-spined stickleback** – only 10 cm (4 inches) long – is scaleless, but has bony plates on its sides. It feeds on worms, crustaceans, molluscs, insects and vegetation. In the breeding season males develop a bright red belly. The female lays her eggs in a nest of weeds built by the male. The eggs are guarded by the male stickleback who fans water over them until they hatch. If young sticklebacks swim away, the male chases them, sucks them into his mouth and spits them back into the nest.

3 **Trout** have been introduced into seas, lakes and rivers worldwide. There are two kinds: the sea trout and the freshwater brown trout. They are alike physically, but sea trout have silvery scales and brown trout have numerous dark spots. **Rainbow trout** are farmed in large quantities. They feed mainly on insect larvae, molluscs and crustaceans. In the wild, female rainbow trout make a nest in the gravel bottom of shallow streams. She deposits her eggs which are fertilised by the male and then covered over. Adult rainbow trout grow to 1 m (3¼ feet). The migratory form of the rainbow trout is known as the steelhead.

4 **Pike** feed on fish, ducklings and sometimes voles. They live in lakes and quiet rivers in the northern hemisphere. A pike's huge mouth has straight lower teeth for seizing its prey and backward-pointing upper teeth to prevent its prey escaping. It is the fastest freshwater fish, capable of speeds over 30 km/h (20 mph). The **northern pike** lurks among vegetation, camouflaged by its mottled colouring, before shooting out to trap its prey. Females are larger than males and some weigh over 23 kg (50 lb). In early spring they lay thousands of eggs over plants at the water's edge.

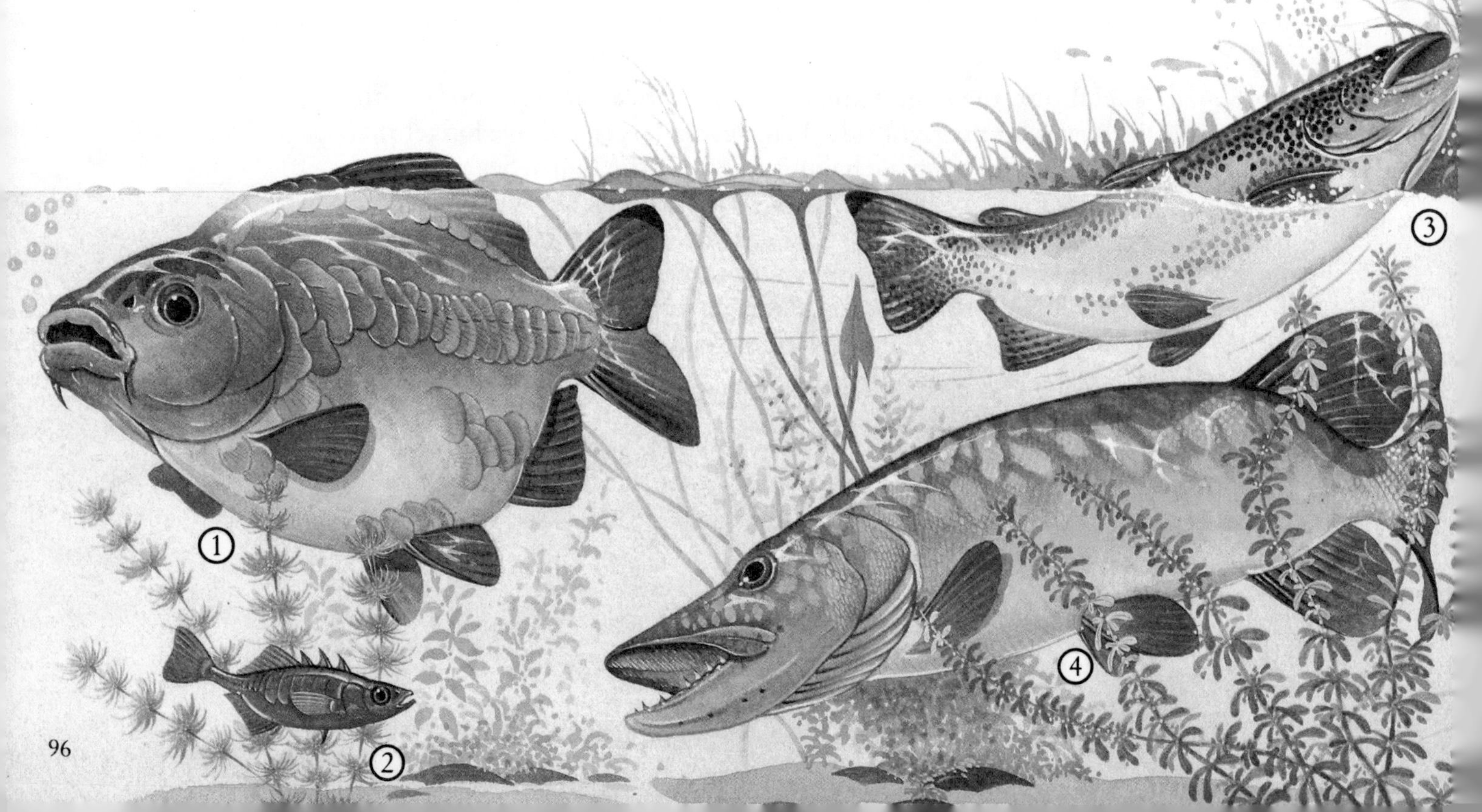

5 **Perch** are found in European lakes, ponds and slow rivers and have been introduced in Australia, New Zealand and South Africa. Young perch gather in schools near the river bed. Their barred markings help to camouflage them as they wait to prey on smaller fish. Larger perch live among aquatic vegetation in deeper waters. They have a characteristic black mark at the end of their spiny dorsal fin. Adults are 51 cm (20 inches) long and weigh about 2 kg (4½ lb). Nile perches are the heaviest of the perches, weighing up to 80 kg (176 lb).

6 **Dace** live in rivers and streams in northern Europe and Asia. They like moderate currents and clean, shallow water. Large schools of dace feed on some plants, insects and their larvae, spiders and other invertebrates which fall into the water. A large adult dace is 30 cm (1 foot) long and can swim over short distances at speeds up to 15 km/h (9.5 mph). Females spawn in spring and their eggs lodge among the gravel until they hatch 25 days later.

7 **Salmon** are migratory fish which travel thousands of miles to return to the river in which they hatched. Memories of the taste of the water draw them on as they leap up waterfalls. Males develop fangs and fight each other; females lay 14,000 eggs in a scrape in the gravel. Many adults then die. When the eggs hatch in spring, the 'fry' stay in the river for 2–3 years, growing rapidly. When they reach the 'smolt' stage they migrate down river to the sea. The **Atlantic salmon** is found in the North Atlantic Ocean where it feeds on shrimps and other fish. It grows to a length of 1.5 m (5 feet), weighs up to 50 kg (110 lb) and can reach a speed of 40 km/h (24 mph).

◀ **Portuguese man o'war** This creature can give a painful sting to a swimmer. Its trailing tentacles are up to 1.5 m (5 feet) long.
A Portuguese man o'war is actually a collection of different parts which can only exist together. Each part has its own job: some capture prey (usually fish); some digest food; others reproduce; and one is a gas-filled bag. This bladder is about 15 cm (6 inches) long. As it floats on the surface, it drifts with the ocean currents. Sometimes storms drive masses of them on to beaches where they soon dry and shrivel.

▶ **Seahorses** are the slowest-moving marine fish. They swim upright in the water. Waving the fin on its back, the seahorse is driven forward. The female places her eggs inside a pouch on the male's stomach. The pouch has a compartment for each baby. Between four and six weeks later, the pouch opens and the young swim free. The male helps them escape by rubbing his belly against a rock. The **spotted seahorse** is 30 cm (1 foot) long and has a prehensile tail. It is found in the Indian and Pacific Oceans.

▼ **Squid** This invertebrate belongs to a group of sea creatures called cephalopods. Their 10 tentacles are attached to the head. The squid squirts a cloud of liquid into the water and moves quickly to escape from a predator. This streamlined swimmer moves by jet propulsion, at speeds up to 51 km/h (32 mph). Water is forced out through a siphon that it points forward, backward or sideways. The **European squid** is 30 cm (1 foot) long. It grabs its prey, mainly fish, with its two longer arms and tears off meat with a hard beak.

▼ **Octopus** This cephalopod has eight arms and is found in warm seas around the world. It has a larger brain for its size than any other invertebrate. An octopus waits in hiding places on the sea bottom to ambush shellfish or any prey it can kill with its poisonous bite. Its long, flexible arms have rows of suckers. When threatened, an octopus squirts out a cloud of black ink and swims backward. Females lay up to 150,000 eggs. The record size and weight for a **common Pacific octopus** is 6 m (20 feet) across its arms and 50 kg (100 lb).

## Crustaceans

Crustaceans are a kind of shellfish, that is invertebrate sea creatures which breathe through gills and have hard, close-fitting outer shells. Lobsters, shrimps and crabs are all crustaceans.

▶ **Soldier Crab** Like other crabs, the soldier crab has a segmented body, four pairs of jointed limbs and a tough shell. Its shell does not grow so it has to cast it off and replace it with a larger one. It uses the pair of antennae in front of its mouth as feelers. It sifts through mud for food and when the tide comes in it digs an airtight chamber to stay in until low tide.

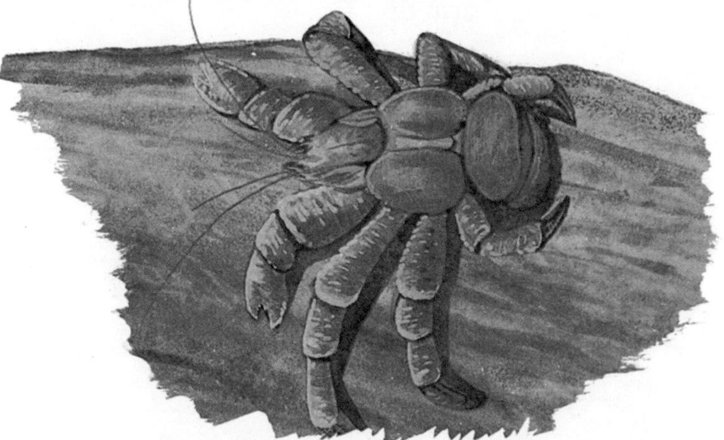

◀ **Coconut crab** This huge crustacean lives on islands in the Indian and Pacific Oceans. Its body length is 30 cm (1 foot). The coconut crab lives in burrows and holes in trees. It is an agile climber. Its pincers are so powerful that it can cut coconuts from palm trees and open them when they are on the ground. Vegetation and dead animals are also part of its diet. Coconut crabs are killed as food for humans. They are now an endangered species.

▼ **Fiddler crab** The male uses its enlarged red claw as a flag, waving it at the female to impress her. If ever the big claw breaks off, the smaller one gets bigger. Fiddler crabs live on mudflats, picking up lumps of mud with pincers and scraping off tiny organisms with a set of hair-fringed blades in front of the mouth. Fiddler crabs are found in Central America and on Pacific islands.

▼ **Hermit crab** This scavenger has a soft lower body. To protect this it crawls into the empty shell of another shellfish or snail. It then carries this 'borrowed' home around with it for protection until it grows too big for it and has to move into a new shell. Some hermit crabs will fight others of their own species to gain a shell. Most live at the bottom of the sea. Some share a shell with a *sea anemone*. Hermit crabs are about 6 cm (2½ inches) long.

1 **Barnacles** are small arthropods which attach themselves to rocks, ships or sea animals. They have a shell with a hinged cover. This protects the body and retains water. When the tide rises, the cover opens, six pairs of feathery legs extend and sweep through the water searching for plankton. Barnacle larvae have a single eye and swimming 'limbs' to propel them through the water. Many of the 800 species of barnacle are no bigger than 1.25 cm (½ inch).

2 **Mussels** are soft-bodied invertebrates with two hinged shells and a mantle (fold of skin) surrounding the body. These bivalve molluscs are slow moving and clamp themselves to rocks by bunches of sticky threads. When the tide is in, their shells open slightly and their gills strain tiny food particles from the water. **Common mussels** – 7.5 cm (3 inches) long – have dark-blue shells and are often found on the same rocks as *barnacles.* They are eaten by *starfish.*

3 **Starfish** are echinoderms (marine animals with symmetrical bodies). Some have as many as 40 arms; most have five. If it loses an arm, another grows in its place. A starfish eats by pulling open the shell of its victim, a mussel or a clam, with its arms. It then pushes its bag-like stomach out of its mouth and on to the soft parts of the victim. It dissolves and absorbs food into its stomach. A **common starfish** is 15 cm (6 inches) from arm tip to arm tip.

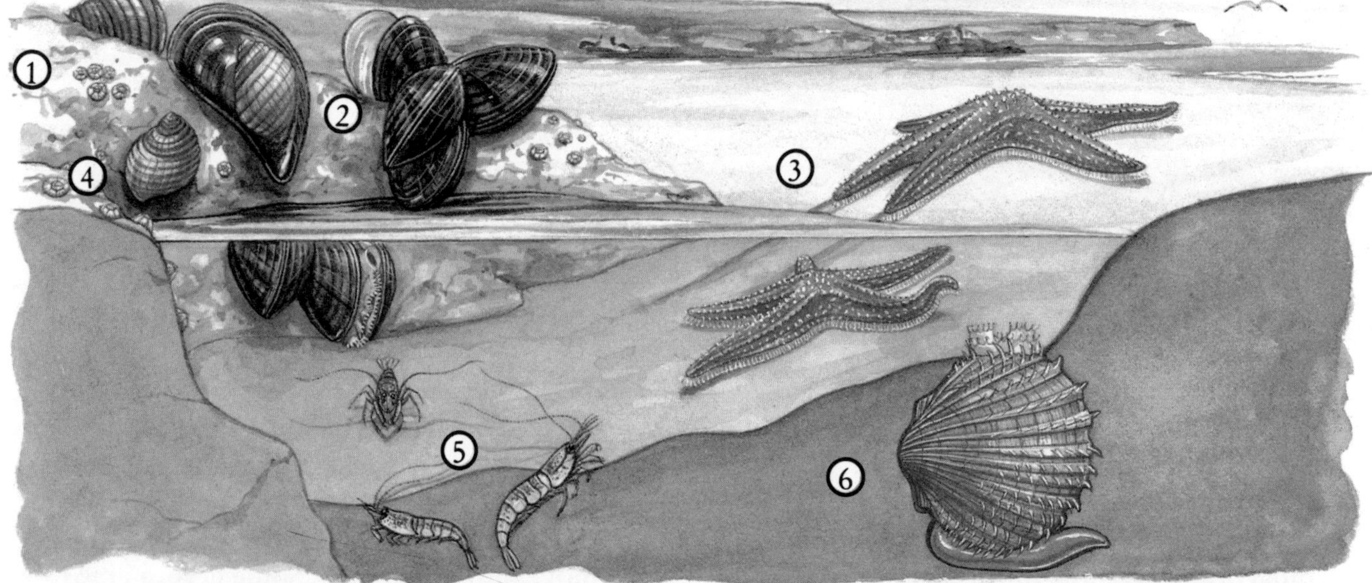


4 **Dog whelks** use their rough radula (tongue) to drill through the shells of mussels, scrape out the soft flesh inside and carry food into their mouths. To get at the edible parts of a barnacle, the dog whelk forces the shell open. Dog whelks are only 3 cm (1¼ inches) high. Their thick shell is usually off-white or yellow, but may be banded. They are found in great quantities on all rocky shores. Their eggs are placed on the underside of rocks.

5 **Shrimps** are crustaceans. They breathe through gills and dig burrows for themselves. They sit and snatch at passing particles with their pairs of hooked antennae. The **common shrimp** is about 6 cm (2½ inches) long. The banded coral shrimp – about 7.5 cm (3 inches) long – lives in coral reefs, picking parasites off the bodies of fish with its claws. The Texas blind shrimp lives at the bottom of lakes in dark caves.

6 **Cockles** are molluscs with a thin, heart-shaped shell. There are about 20 ribs on the shell. The colour range is from yellow-brown to red. A cockle burrows into mud and puts up two short, fleshy tubes which suck in food particles from the water. It is found on European seashores. The **spiny cockle** is 5 cm (2 inches) long. Cockles are attacked and eaten by *starfish.* People also dig them up and sell them as seafood.

# INSECTS

There are about one million living species of insect and new ones are being discovered all the time. Insects are invertebrates which breathe through air tubes. An insect's body is divided into three parts – head, thorax and abdomen. On its head is a pair of antennae, sensitive to both scent and vibration, and its eyes and mouth parts. Their compound eyes consist of hundreds of tiny beads. Three pairs of legs and one or two pairs of wings are attached to the thorax. An insect goes through changes of structure during its life which is called its metamorphosis. As it grows, it casts off an outgrown skeleton and a larger one hardens in its place. Each casting is called a 'moult'. Most insects start as eggs from which larvae hatch. Many insects have developed amazing methods of protection against predators. Included in this section are some creatures which are not strictly insects, such as spiders, scorpions and worms.

► **Honey ants** are found in Australia. Among honey ant colonies there are special workers known as repletes or honeypots. These workers are pumped so full of sweet juices by other workers that their bodies become bloated and look like transparent balloons. Special chambers are prepared in the honey ants' nest and the swollen honeypots hang upside down from the roof, motionless. They wait to be 'tapped' by other honey ants when no more food can be found outside. Some primitive people hunt for honey ants' nests and dig honeypots out of the ground to extract their honey.

◄ **Driver ants,** also called safari ants, are wandering hunters in Africa. Like *army ants* in America, they are flesh eaters. Thousands of these blind ants march in a single colony. The queen ant is half carried, half dragged by the workers. They are flanked by soldier ants as they move up to 70 m (230 feet) in one night, killing and eating every living creature they come across – insects, reptiles and mammals. It is impossible to stop them, unless they are sprayed with a powerful insecticide. The only warning they give is a curious rustling noise made by the driver ants' hard bodies as they run along.

▼ **Army ants,** also called legionary ants, are the American version of *driver ants*. The queen ant is 5 cm (2 inches) long with a swollen body but no wings. A colony of up to half a million army ants forage for weeks, day after day. They eat other insects, worms and molluscs. Soldier army ants have very powerful jaws and once they have hold of a victim they do not let go until they have torn off a piece of flesh. The queen lays up to 120,000 eggs a week.

◀ **Ant-lion** This insect is so called because it feeds mainly on ants. Also known as a doodle bug, the ant-lion larva digs a pit in dry or sandy soil. It does this by going round and round backwards in small circles, tossing away the sand with upward jerks of its head. Then it buries itself with its jaws wide open, waiting for its prey. When a victim falls down the slippery sides into the pit, the ant-lion larva paralyses it with a bite and sucks the body dry. The larva stage lasts for up to three years. Adults live only long enough to mate and lay eggs. Some adults are 7.5 cm (3 inches) long.

▶ **Weaver ants** are found in Africa, India and Australia. They construct nests by sewing leaves together. Sometimes workers form an ant chain to bridge the gap between two leaves. Each ant holds the waist of the ant above it. They pull the two leaves slowly together. Then other workers move in, holding young larvae in their jaws. Gently squeezing the larvae to produce silk, they move back and forth until the edges of the leaves are joined by silken fabric. Silk is also used to line the nest. The larvae have no silk left when the time comes to form a cocoon so the naked pupae are sheltered in the silk-lined nest. Weaver ants feed on caterpillars, bugs and beetles.

▼ **Leaf-cutting ants**, also called 'parasol ants', use their scissor jaws to shear green leaves into sections which they carry to their nest. These Central and South American ants build huge nests underground – up to 6 m (6½ yards) deep. Seven million ants live in a single nest. They chew the plant material, mixing it with saliva, to form a compost. They feed on the fungus which grows in the compost.

▶ **Termites** construct nests with many rooms inside. Some live underground, others build mounds out of wood pulp or soil mixed with saliva. Each mound has a kind of air conditioning. Air ducts in the outer ridges are opened or closed to keep a steady temperature. **African termites** build mounds up to 12.8 m (42 feet) high. There are 1900 species of termite. The queen of the largest species is 14 cm (5½ inches) long and produces up to 30,000 eggs a day. Worker termites are small, blind and wingless. They build the nest, bring food to the queen and care for her eggs. Soldier termites defend the colony, beating their heads on passage walls to sound the alarm.

◀ **Froghopper** This bug is known as the 'spittle insect' in the USA. A froghopper will take off in a flying leap, when it is approached. Females lay eggs in October or November and these hatch early in spring. The young (nymphs) crawl down the plant stem, find a suitable site and begin to suck sap from inside the stem. To protect itself from the sun and from enemies, such as birds, the young froghopper creates a mass of white froth. This is formed by mucus from the abdomen blown into bubbles by air. When the adult emerges after metamorphosis, it hops and leaps about, feeding on plants and shrubs. Adults are rarely over 13 mm (½ inch) long.

▶ **Ladybird** These well-known beetles live in most parts of the world. In the USA they are known as ladybugs. They have short antennae and their legs have only three joints. Their bright colours and spots warn birds that they taste unpleasant if eaten. They do much good by feeding on aphids which damage plants. The **seven-spot ladybird** is found in Europe. Its larva is about 8 mm (⅓ inch) long and the adult is slightly smaller. Large numbers of ladybirds gather in dry crevices to hibernate. Some species sometimes fly up mountains to huddle together under snow. Many of them perish from the cold or are eaten by other insects, birds or bears.

◀ **Hercules beetle** This is one of the world's largest beetles and is found in South American rain forests, Central America and the West Indies. Male hercules beetles are up to 20 cm (8 inches) long. More than half of this length is taken up by the two horns. The upper horn is an extension of the thorax and the head extends as the lower horn. By making up and down movements of his head, the male works the two horns like a pair of forceps. They are used only at mating time, as males fight each other.

▶ **Bombadier beetle** This European species has long legs. It defends itself in a very clever way. Two sacs at the rear of the bombadier beetle's abdomen contain a smelly fluid that forms a smoke-like gas on contact with air. It can eject a series of four or five shots, each with a 'pop' sound. The beetle then escapes as its predator chokes on the gas. Reptiles and birds spit out a bombadier beetle as soon as its fluid is in contact with their mouth. Adult bombadier beetles are about 8 mm (⅓ inch) long. They live under stones at the edge of fields in warm, dry areas. Often several of them are found under the same stone.

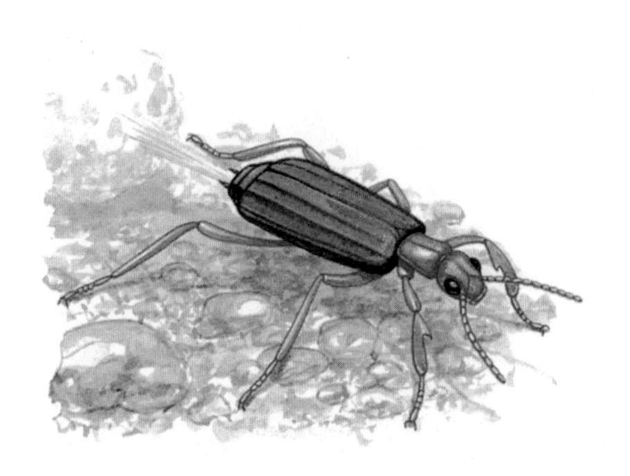

◀ **Centipede** Although its name means '100 legs', the actual number of pairs of feet ranges from 15 to 177, depending on the species. There is one pair of legs for each body segment. The legs of the first segment behind the head are modified into poison claws. Centipedes feed at night on any animal of a suitable size, including worms, slugs and other insects. In America giant centipedes are 25 cm (10 inches) long. This European species is 2.5 cm (1 inch) in length. The female carries her eggs between her hind legs before placing them in the soil. Sometimes she rolls them on the ground to give them a protective covering of soil.

▶ **Cockroach** There are 3500 species of cockroach around the world. The cockroach's long legs are well adapted for running. Its flattened body enables it to squeeze through tiny cracks. Most species fly well, attracted to light at night. When at rest, the wings are folded back. Cockroaches are scavengers, feeding on any remains they come across. Their eggs are enclosed in a bean-shaped, flat case. When they hatch, worm-like grubs emerge and moult their outer skin. These young cockroaches look more and more like their adults with each moult. The **American cockroach** is found in the wild in Florida. It is up to 5 cm (2 inches) long.

◀ **Scorpion** This carnivore is not a true insect. Like the spider, it has eight legs and is an arachnid. Scorpions ambush prey before paralysing or killing it with poison from the sting on the end of its tail. The most dangerous species is the fan-tailed scorpion of Africa whose venom can kill a dog in seven minutes. The largest scorpion species is found in India: males are 18 cm (7 inches) from tip of pincers to end of sting. The **Sahara scorpion** in Africa is half this length. This scorpion escapes the heat of the sun by digging deep burrows with its claws. Eggs hatch inside the female's pouch, the young crawl out and clamber up on to her back.

▶ **Earwig** This insect is common in Europe and most warm areas. Earwigs feed at night on caterpillars and flower petals. Some earwigs secrete a nasty odour to protect themselves from attack. Female **common earwigs** are 2.5 cm (1 inch) long. Males are smaller. Though the female of this species has wings, she rarely flies. Adults hibernate underground in winter. The female lays 40–50 eggs in the soil in spring. She watches over her young (nymphs) until they can find their own food. She raises her rear pincers to scare enemies away. After guarding her brood, she dies.

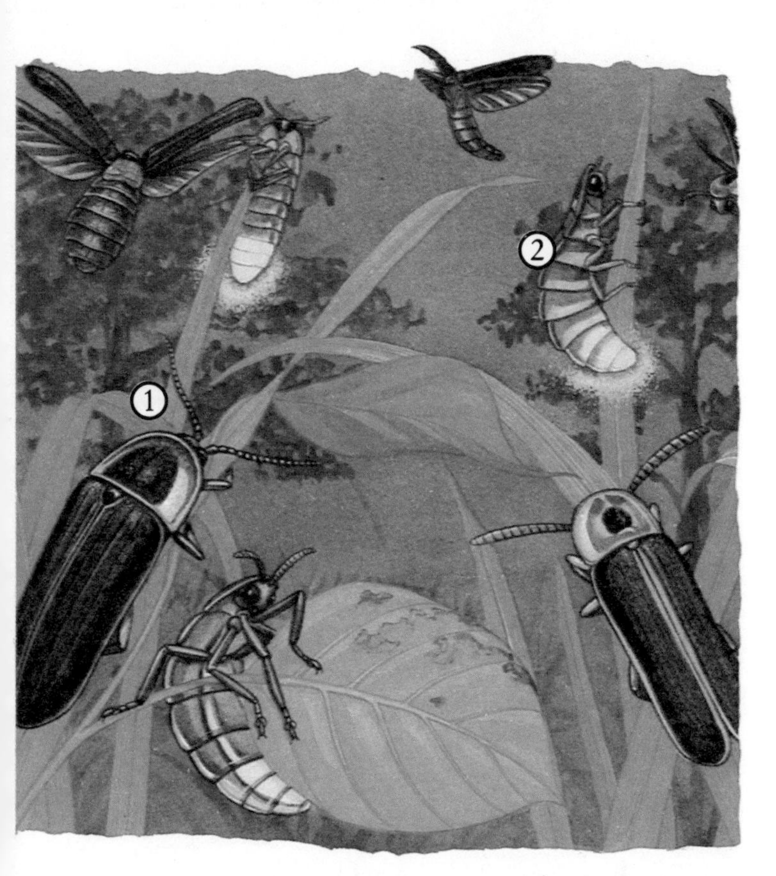


◀ 1 **Fireflies** and 2 **glow-worms** are really beetles. They produce a light at night. Male and female fireflies flash lights to attract one another. Light organs in the abdomen produce chemicals that shine when mixed with oxygen. Only male fireflies have wings and can fly. Most species of glow-worm are tropical fireflies. The wingless female – only 16 mm (⅔ inch) long – is larger than the male. She has a soft, greenish light in the last segments of her abdomen. Glow-worms hide under stones or rubbish by day. At night the male flies around looking for a mate. The female will switch her light off if disturbed. Firefly and glow-worm larvae feed on snails and slugs. They pierce the shells with their mandibles (jaws) and inject venom to paralyse the creature. Adults of many species do not eat and those that do, feed only on nectar and pollen.

▶ **Flea** The adult flea is a parasite, about the size of a pinhead. Its mouth parts are modified for piercing skin and sucking blood. Many flea species can jump. One species jumps 30 cm (1 foot) in a single leap. Usually fleas walk about, gripping hair or feathers with strong claws. Wherever dirt collects, fleas breed. Their white larvae have no legs or eyes and feed in the dirt. When the adult flea emerges, it attaches itself to the nearest host, which may be a cat, dog, rat, bird or a human.

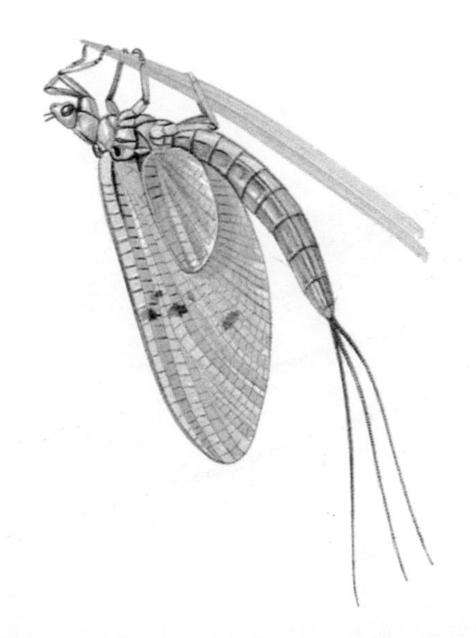

◀ **Mayfly** The larvae of this insect live about two years in ponds, feeding on small particles. On metamorphosis, the mayfly emerges from the water as a winged but immature adult. The next day it moults again to produce the fully grown adult. These adults eat no food and their guts contain only air. Their four wings are transparent and cannot be folded back when resting. The adults lay their eggs and die – one day after maturing. This female European mayfly is 2.5 cm (1 inch) long.

► **Stick insect** There are 2000 species of stick insect and its near relative the leaf insect. Camouflage, in terms of shape and colour, is an important defence for these slow-moving insects. Stick insects are brown or green making them difficult to see as they sit motionless in trees or bushes during the day. They feed on plants at night when predators cannot see them moving. Adults of the European species have no wings and are 6 cm (2½ inches) long. Their eggs have hard shells and look like seeds. This is also useful disguise as it probably saves them from being eaten by animals that feed on insect eggs.

◄ **Treehopper** This insect belongs to a group of bugs found mainly on trees in tropical areas. This species is known as the **thorn bug** because it look like the thorn on the stem of the plant on which it lives. This disguise is excellent protection from predators. The young (nymphs) do not have the same horn-like projection as the adult. Instead the abdomen is long and pointed. Adult thorn bugs are about 8 mm (⅓ inch) long.

► **Praying mantis** This usually pale green insect is found in western Europe and eastern America. It is well camouflaged as it sits motionless among vegetation. A praying mantis feeds on flies, grasshoppers, butterflies and other insects. It shoots out its front legs at great speed and snaps them shut around its prey. The spines on its forelegs hold the insect firmly while the mantis eats it with its strong, sharp mandibles. Females of this species are 6 cm (2½ inches) long – much longer than males. After mating, a female mantis often eats her male partner. She lays her eggs in a frothy case.

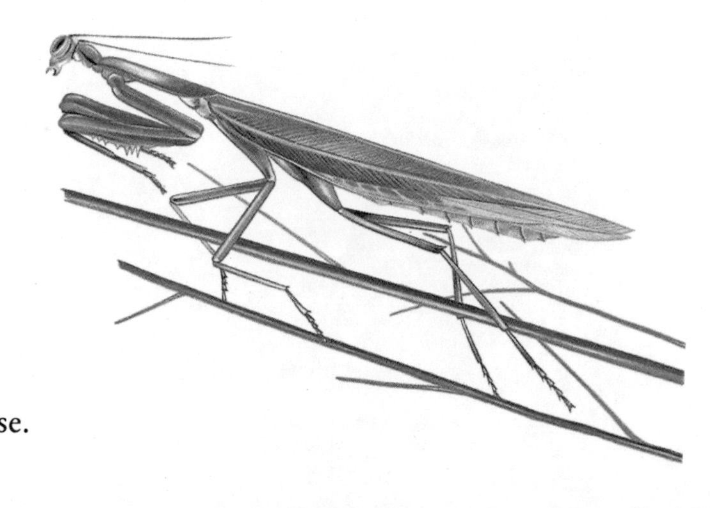

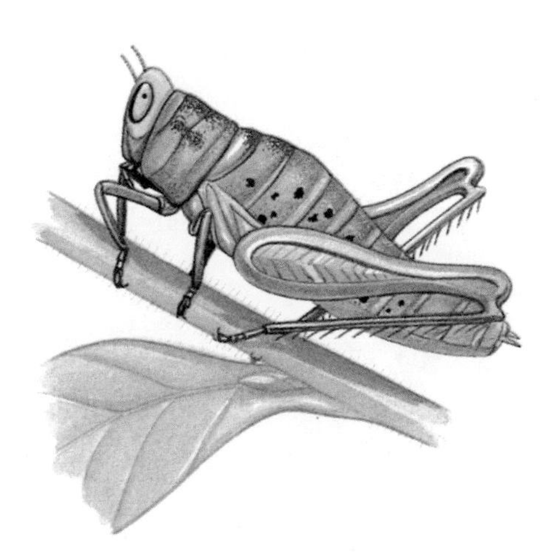

◀ **Grasshopper** As its name suggests, this grass-eating insect moves by hopping. It has well-developed hind legs and strong claws. The grasshopper's 'song' is produced by stridulation. This is the rubbing of one part of the body (file) over another part (scraper). The file is a leg and the scraper is a wing. By moving its leg very rapidly the grasshopper produces pulses of sound. Its ears are at the base of its abdomen. If a grasshopper damages a leg, it eats it and another grows in its place. Females lay eggs undergound, in crevices or plant stems. This **African grasshopper** is brightly coloured. Most species are pale green in colour.

▶ **Bush cricket** This insect's chirping 'song' is often heard coming from bushes, but the source of the sound is hidden from view. A cricket 'sings' by raising its wings and rubbing them together. The left wing (file) lies on top of the right wing (scraper). A cricket hears with its legs. The earholes are near the knee on the front pair of legs. Bush crickets were called long-horned grasshoppers because of their very long antennae. The **oak bush cricket** lives on oak trees. It crawls about at night, feeding on foliage. Males of this species do not stridulate but attract females by drumming with their hind feet on an oak leaf. They are about 13 mm (½ inch) long.

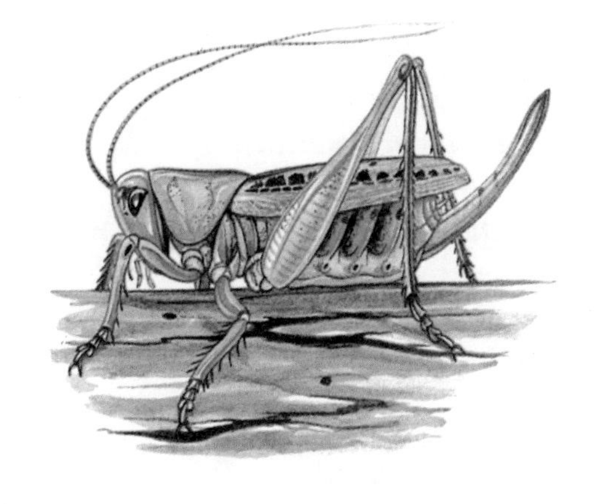

◀ **Mole cricket** This insect has short front wings, but fully developed hind wings. It flies on warm summer nights and the male 'sings'. A mole cricket spends most of its time underground. Its enlarged front legs are adapted for burrowing. Like the *mole*, this cricket makes shallow tunnels for long distances beneath the surface of the soil. As it digs, it feeds on tender roots. Its mandibles are very efficient shears for nipping through rootlets. The mole cricket is brown and is covered in fine hairs. Adults are about 5 cm (2 inches) long. They are now rare, possibly due to human activities such as drainage of the soil.

## Spiders

Spiders are not insects. Like scorpions, they are members of the arachnid class. They have eight legs and, in most cases are aggressive carnivores: their diet is made up of small creatures they trap and eat. All spiders spin silk which is used either to hold eggs or to trap food in webs. Male spiders are often in danger of being eaten by the female after mating. The male of one species offers the female an insect wrapped in silk as a present before they mate, so that she will eat this rather than him.

▶ **Wolf spiders** are found in dry areas. They do not catch their prey in webs but chase after and jump on it, stabbing the victim with poison claws. The grey-coloured male, only 5 mm (1/5 inch) long is smaller than the female, 13 mm (1/2 inch). After laying her eggs on a web placed on the ground, the female rolls this up and fixes the cocoon on her abdomen. Before the eggs hatch, the adult spins a web around herself and her eggs. Once they have hatched, the young spiders cluster on the female's back. She carries them round until they are one week old.

▼ **Bird-eating spider** This arachnid stalks its prey silently and kills it with a poisonous bite. It does not usually kill birds, but prefers insects, mice and small lizards. The **Guyanan bird-eating spider** is found in South America. It has a legspan of 25 cm (10 inches). Its body and legs are covered in thick hairs. This spider often has a bald patch where it has removed hair from its hind legs and thrown it at an attacker. This form of defence is effective if the predator chokes on the hair.

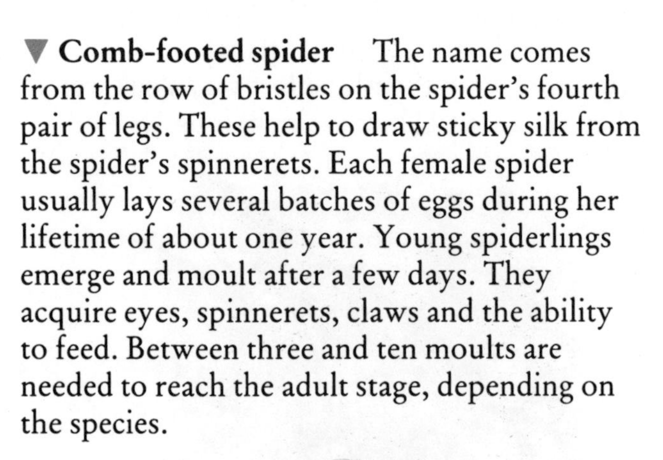

▼ **Comb-footed spider** The name comes from the row of bristles on the spider's fourth pair of legs. These help to draw sticky silk from the spider's spinnerets. Each female spider usually lays several batches of eggs during her lifetime of about one year. Young spiderlings emerge and moult after a few days. They acquire eyes, spinnerets, claws and the ability to feed. Between three and ten moults are needed to reach the adult stage, depending on the species.

▶ **Black widow spider** This is the most dangerous arachnid. The female's venom is very powerful, though rarely fatal to humans. A female black widow spider is 13 mm (½ inch) long and has a red marking on its body. The male is timid, harmless and only 6 mm (¼ inch) long. It only bites when frightened. If a female black widow is hungry she may eat a male. They are the shortest-lived of all spiders. Males live for an average of 100 days and females survive for about 270 days. Black widow spiders are found in America, the West Indies and around the Mediterranean.

◀ **Web-throwing spider** This African spider spends the daytime in a twig-like posture. At night it weaves a web about the size of a stamp. It is made of a special silk which is not sticky but is stretchy. A web-throwing spider holds the web between its legs and waits about 5 cm (2 inches) above the ground. When an insect comes along, the spider stretches the web to six times its first size and throws it over the prey. Then it bites the victim and wraps it in silk. It carries the package to a nearby twig and slowly begins to feed.

▼ **Crab spider** This common arachnid does not build a web. Instead it lies in wait and seizes passing insects with its long front legs. The crab spider does not chew its food, but bites its victim and sucks out the contents. Some species can change colour to match the plant or ground so closely that they are almost invisible. Crab spiders get their name from their habit of moving sideways. Females are 8 mm (⅓ inch) long and the short-sighted males are shorter.

▼ **Silk-throwing spider** This arachnid lives in Australia. It uses a special trick to catch its prey. The silk-throwing spider spins a single silken thread and places a drop of glue on the end. Then it lies in wait. When an insect approaches, the spider whirls its sticky thread round and round and throws it like a lasso. The prey is unable to escape. As its victim struggles, the spider descends to the ground to capture it.

▶ **Mosquito** This dangerous insect is the carrier of diseases such as malaria and yellow fever. There are 2000 species of mosquito throughout the world, including the Arctic. Only the female mosquito bites. She pierces the victim's skin with her long proboscis (mouth parts) and sucks blood, which is needed for her eggs to develop. Then she lays her eggs on or near the surface of water. After hatching, the larvae hang by their breathing tubes from the surface. A group of bristles round the mouth help to filter food. They next go through the pupal stage, after which, as adults, they mate and then die. This adult female – 8 mm (⅓ inch) long – is found near water in Europe.

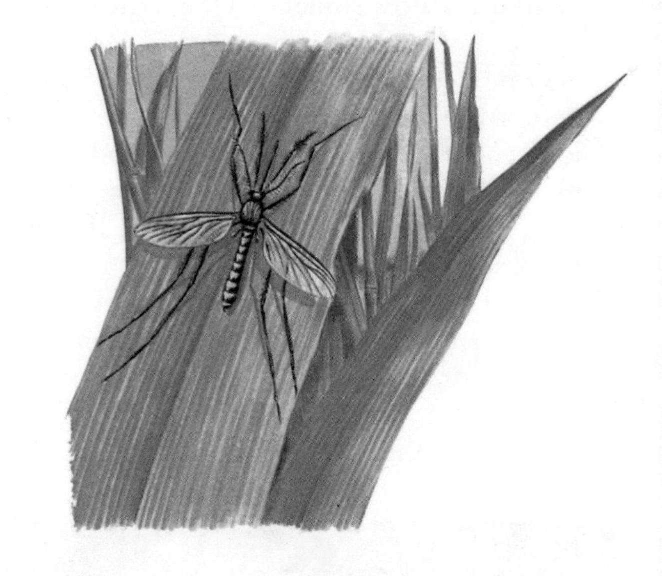

◀ **Housefly** There are around 70,000 known species of true fly. The common housefly is a dangerous insect because it can transmit many diseases to humans, including leprosy, smallpox and the bubonic plague. The housefly breeds in composts and refuse tips. Larvae are legless maggots and take about three weeks to mature. They feed on rotting vegetable matter. The adult housefly enters houses and walks all over food. Its mouthparts soak up juices like a sponge. Millions of germs may be present on a single fly. An adult housefly is about 8 mm (⅓ inch) long and has only two wings. It can taste food with its feet.

▶ **Bengali fly** This Asian species feeds on the eggs of *driver ants*. The fly acts like a divebomber. It hovers over a column of driver ants on the march, as they carry their eggs. Then it drops down on to the ants suddenly. A tug-of-war follows between an ant and the fly with the egg pulled back and forth. As soon as the driver ant puts down its egg, the Bengali fly snatches it and flies away. This fly is slightly larger than the *housefly*.

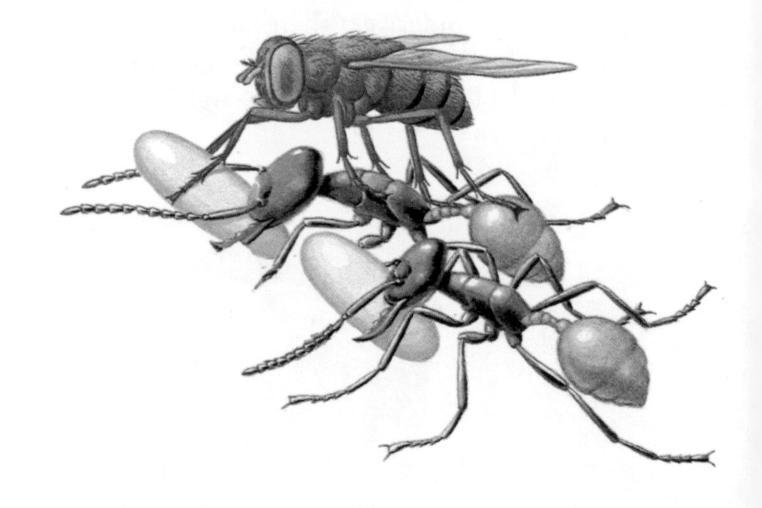

## Wasps and bees

Wasps and bees have two pairs of wings. The forewings are linked to the hind wings. Their mouthparts are adapted for biting, lapping and sucking. They have a waist between the thorax and abdomen. Females seek food for their larvae.

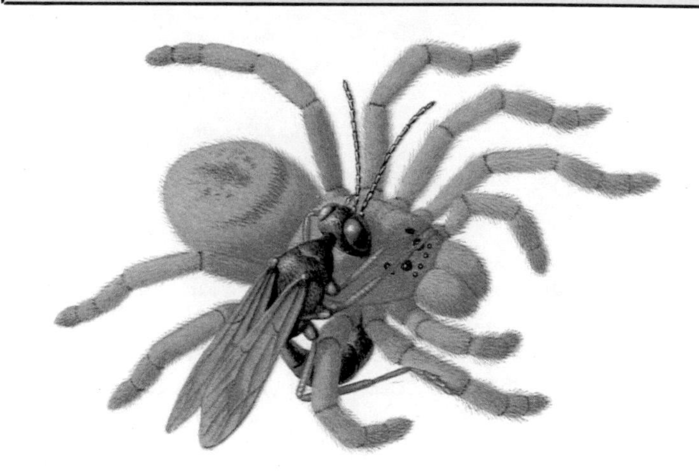

◀ **Tarantula hawk** In South America the female lures a tarantula spider out of its nest. A fierce battle follows. If the tarantula wins, it feasts itself on the wasp. If the wasp succeeds in stinging the spider, its venom paralyses the prey. The tarantula hawk wasp drags the spider for distances up to 50 m (164 feet) to its own burrow. She lays a single large egg on the spider and covers it up. The larva hatches in a few days and feeds on the spider. This wasp is 6 cm (2½ inches) long.

▶ **Potter wasp** The female potter wasp provides food for her young. She either builds a nest of clay (as shown here in cross-section) or burrows in wood or soil. The cell is stocked with caterpillars which are paralysed by the wasp's sting. One egg is laid inside before the top is sealed. When the larva hatches, it feeds on the caterpillars. The potter wasp has yellow legs and a thin additional section between its thorax and abdomen.

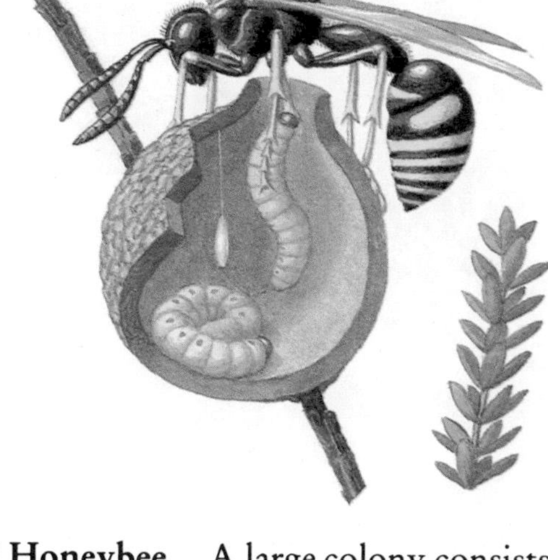

◀ **Honeybee** A large colony consists of a queen, about 200 males and some 70,000 workers. A suitable nesting site is often a hollow tree. A queen's first batch of eggs are worker bees. The workers build a honeycomb of six-sided cells. They produce honey from nectar they gather from flowers and store it in the cells. They also collect pollen to feed larvae from which the young queens are reared. A bee draws heat from the sun. Its thorax muscles have to be warm to lift the bee into the air. A worker bee is about 16 mm (⅔ inch) long.

## Butterflies and moths

Butterflies are usually brightly coloured and active during the day. Moths tend to be duller and active at night. Both have two pairs of wings and they follow the same life-cycle. There are four stages: egg, larva, pupa and imago (adult). The larva (caterpillar) is very different in appearance from the adult form. As the larva grows, it sheds its outer skin. It either spins a silk cocoon or its skin hardens. During the pupal stage the body is broken down and reformed as an adult butterfly or moth.

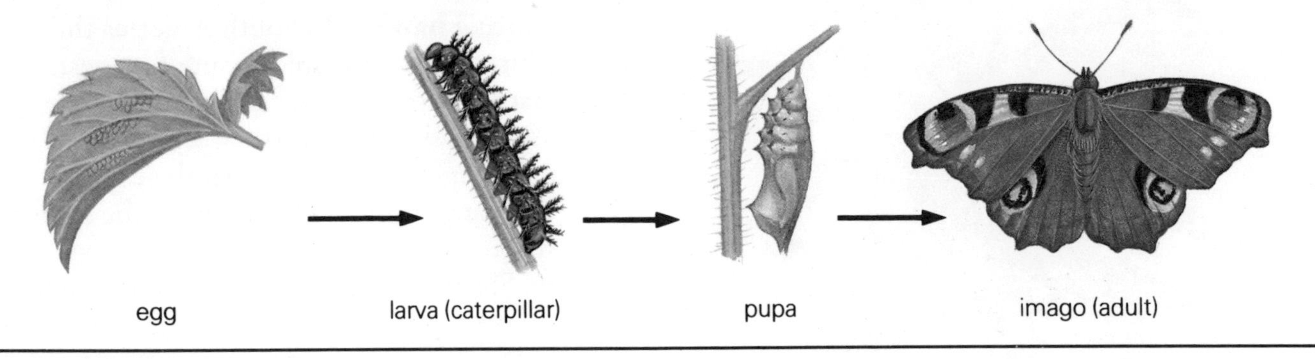

◀ **Swallowtail** This butterfly is found throughout the world except in the Arctic. It is becoming rare because its natural habitat, the countryside, is being destroyed. A female swallowtail has a wingspan of 10 cm (4 inches). She lays greenish eggs which stick to the lower surface of leaves. The caterpillar that hatches is yellow with black blobs and has two fat horns at the front. When winter comes, the swallowtail caterpillar attaches itself to a plant and spins a cocoon. It spends winter as the pupa. In the spring the swallowtail emerges.

▶ **Dead-leaf butterfly** Found in India, China and Australia, this butterfly is 7.5 cm (3 inches) long. The upper side of its wings is brightly coloured, making it easy to spot when in flight. When it lands on a stem or twig, it claps its wings together over its back. The shape and position of the dead-leaf butterfly's wings imitate a leaf. The underside of its wings is patterned like the veins of a leaf.

◀ **Hawk moths** These large moths have long, narrow wings. They are swift fliers. A male can detect the scent of a female 5 m (16 feet) away with its antennae. The **striped morning sphinx** (right) has a 7.5 cm (3 inch) wingspan. The **elephant hawk-moth** (left) visits various flowers shortly after sunset. Female hawk-moths lay about 100 eggs from which big caterpillars emerge. There are two pairs of dark spots behind the caterpillar's head, like black eyes. When disturbed, it puffs out its eye-spots to appear threatening.

▶ **Peppered moth** The speckled form of this European moth is found in parks and gardens. It is well camouflaged against tree trunks covered in lichen. This excellent protection makes it hard for bird predators to see it. In smoke-blackened city areas a dark-coloured variety has evolved. Female peppered moths are larger than the males and have thinner antennae. They have a wingspan of 5 cm (2 inches). Their larvae feed on various trees and bushes. They pupate in the soil, spinning a brown cocoon.

◀ **Silk moths** The fine threads of the cocoon spun by these moths are used for making silk. The giant silkworm moth has a wingspan of 25 cm (10 inches). When resting on tree bark, the **Io silk moth** is hard to see. If this North American insect is attacked, it flicks out its front wings. The pattern on its other wings looks like a pair of eyes. This scares away most predators. Io silk moth caterpillars have pointed spurs which irritate animals that attempt to eat them. They feed on the leaves of various trees, including willow, elm and poplar.

▲ **Waterboatman** Also known as the 'back swimmer' because it swims on its boat-shaped back, this water bug is about 13 mm (½ inch) long. It is found in fresh water all over the world, including 4500 m (15,000 feet) up in the Himalayas. It feeds on tadpoles and beetle larvae which it detects by sight and by vibrations in the water. It can inflict a painful sting. This water bug can breathe underwater by using air stored beneath its wings. At night a waterboatman sometimes leaps out of the water and flies around.

▲ **Water spider** This arachnid spends all of its time in ponds, under the water. It builds a kind of diving bell with silk and fills it with air brought from the surface. Inside the bubble the water spider breathes the stored air. When an insect lands on the surface of the water, the spider rushes out, spinning a silken rope. After catching its victim, it hauls itself and its prey along the rope back to its home. Most water spiders are about 13 mm (½ inch) long, but some adults grow to twice that length.

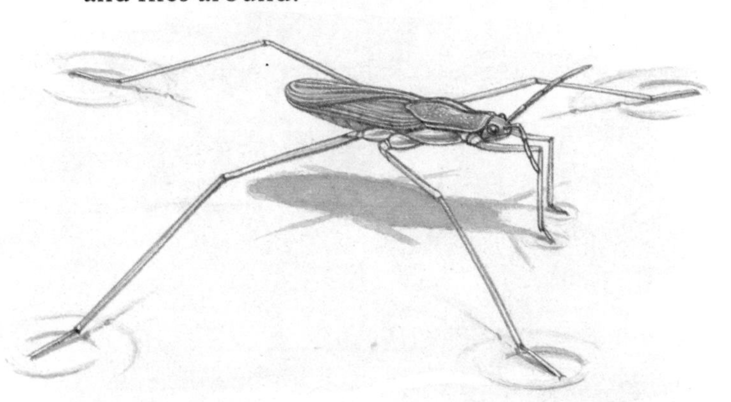

▲ **Pondskaters**, found in Europe, Asia and Africa, are also known as water striders. These 8–10 mm (⅓ inch) insects have wax-coated feet. Their six tiny legs spread out, making tiny dimples on the surface of the water. They use their front legs to grasp their prey, their middle legs as oars to row over the surface and their hind legs as rudders. Pondskater adults seek sheltered places away from water during winter. They return to ponds, lakes and streams in spring and females lay their eggs in May. The young become adults within two months.

▲ **Water snails** rise frequently to the surface to take in fresh supplies of air. They have two long tentacles which cannot be withdrawn. Their eyes are beneath these tentacles. The **great pond snail** is 2.5 cm (1 inch) wide and 5 cm (2 inches) high. Its thin shell ends in a tapering spire. It is mainly herbivorous, feeding on algae which it grates with its radula (tongue-like organ), but it will eat newts. Great pond snails prefer either slow-moving or stagnant streams in Britain. Water snails bear up to 50 live young at a time.

# BIRDS

There are about 8600 species of bird distributed around the world. The common feature is their covering of feathers. These exist in a range of shapes, sizes and colours. They have many functions: from insulation to a means of visual communication. Wing feathers provide the lift and thrust for flight and tail feathers help the bird to steer while in the air. Most birds fly, although a few (e.g. rhea and emu) have lost the ability. Birds range in size from the tiny hummingbird to the ostrich that stands taller than an average human being. They are warm-blooded vertebrates with four limbs, the front pair of which are modified as wings. Their toothless beak is covered with horny plates.

Like mammals, birds are descended from reptiles. Their feathers evolved from the reptiles' scales. Birds have huge breast muscles and a large heart in proportion to their size. Their bones have thin walls and are filled with air. Female birds lay eggs with a tough shell. Each egg contains a large yolk, the food reserve for the developing chick.

▲ **Falcon** All 60 species are daytime birds of prey, often found on mountains or sea cliffs. A **peregrine falcon** has pointed wings. Flying above other birds, it makes a dramatic dive downward, reaching speeds of about 360 km/h (224 mph). The prey is killed outright as the falcon seizes it with its talons (claws). A peregrine is up to 50 cm (1⅔ feet) long. It sometimes feeds on mammals and amphibians. Females lay three or four eggs either on a ledge or on the ground. The number of peregrine falcons is declining as they die from the pesticides consumed by their prey.

▲ **Harpy eagle** A female – 90 cm (3 feet) long and weighing 7.7 kg (17 lb) – is the largest and heaviest eagle. Harpy eagles are found in South America. They are now rare due to the destruction of vast areas of forest and shooting by hunters. Their short wings enable them to chase prey through dense forest. They grab monkeys, sloths, opossums and some birds with their sharp talons. The harpy eagle's nest is a platform of sticks high in the tallest trees. Chicks stay with parents for about 12 months. Harpy eagles probably breed every other year.

► **Golden eagle** This predator has huge, curved claws. It has a hooked bill and feathered legs. Its wingspan is 2.8 m (9⅓ feet). When hunting, a golden eagle soars for long periods searching for prey. It can spot a hare from a distance of 2000 m (6560 feet). Rabbits, hares, foxes and game birds are killed by the eagle's crushing talons. Two eggs are laid in a nest high in a tree or on a ledge. One nest was reported to be 4.6 m (15 feet) deep and was used for 45 years. In many cases the first born chick kills the younger one.

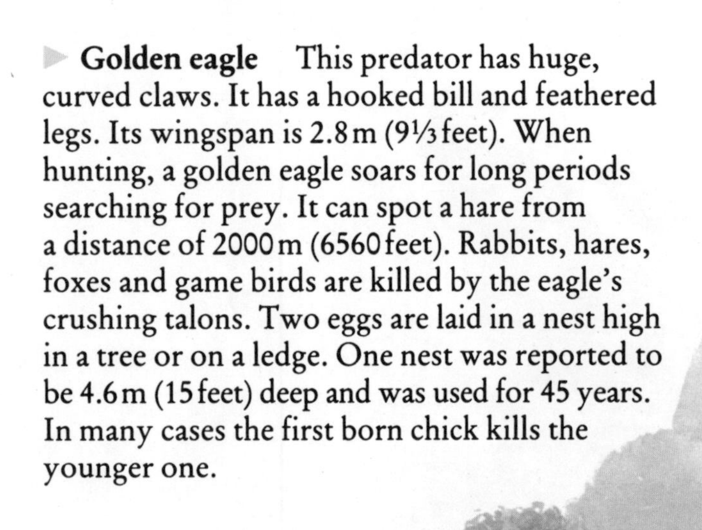

◀ **California condor** This large bird is now endangered because of hunting and destruction of its North American habitat. In 1986 only five pairs were left in the wild. The California condor soars at great heights. It can glide up to 96 km (60 miles) without flapping its wings. The condor does not fly in bad weather or if there is no wind. California condors feed on carrion (dead animals). A single egg is laid every other year on the ground in a cave or a cliff-hole. Both parents feed and tend their young for a year. A young condor does not attain full plumage until it is seven.

▶ **Andean condor** With a wingspan of 3 m (10 feet) and weighing 11 kg (24 lb), this is one of the largest and heaviest flying birds. This relative of the *vulture* has an unfeathered head and neck which enables it to plunge into a messy carcase without soiling its plumage. Up to 20 may be found feeding at a single carcase. Andean condors are found in South America – from sea level to as high as 6035 m (19,800 feet). Females lay one white egg in a nest of sticks and debris. The chick stays with its parents for 16 months. In 1964 an Andean condor died in a Russian zoo aged 72 years.

◀ **Secretary bird** In open country in Africa this bird spends most of its time on the ground walking with long strides. It may cover 30 km (20 miles) each day. The distinctive crest on its head bobs up and down as it walks. A secretary bird runs after its prey and catches it with a swift thrust of its head. It kills larger prey, like snakes and small mammals, by stamping on them. In fact, it will eat almost anything crawling on the ground. A secretary bird chases intruders off its breeding area. It nests at the top of trees.

**Osprey** Also known as a fish-hawk, this bird is found worldwide near rivers, lakes and coasts. It feeds almost exclusively on fish. To help it grip its slippery prey the soles of its feet are studded with small spikes. The osprey hovers briefly, then plunges feet first into the water. It returns to its perch with a fish grasped in both feet. Pairs make a large nest of sticks, seaweed and debris on the ground. The same nest is often repaired and re-used year after year. Three eggs are usually laid. After breeding, northern groups migrate south for winter. Male and female ospreys look alike and are 60 cm (2 feet) long.

**Goshawk** This aggressive bird is capable of killing game birds, rabbits and hares. It weaves in and out of the trees and soars over the tree-tops. The **northern goshawk** is found in northern regions, from North America to Japan. It has a wingspan of 1.2 m (4 feet). Males and females spend the winter alone. In the spring, pairs meet up and give flight displays to one another. Pairs roost together while nest-making and perform a screaming duet each day before sunrise. Usually three eggs are laid by the female. The male brings her food during the incubation period. Chanting goshawks of Africa feed mostly on lizards.

**Vulture** Its feet are adapted for perching rather than for grasping. This scavenger often arrives late at a carcase. Its long bill is ideal for picking out the small pieces of flesh overlooked by larger animals. The **Egyptian vulture** is found in Europe, North Africa, the Middle East and India. It migrates, following animal herds across deserts and cultivated land. The Egyptian vulture is one of the few birds that uses a tool. It breaks ostrich eggs by dropping stones on them from its bill. An adult is 60 cm (2 feet) long and will hiss and growl in anger or if alarmed.

▲ **Barn owl** This nocturnal hunter has soft edges and tips to its wing feathers. When a barn owl swoops down, it makes no sound. Its prey – mainly rodents and birds – do not hear it approaching. A barn owl has a heart-shaped face and a long beak covered in feathers. It is found worldwide, except in Asia. During the day it roosts in farm buildings, hollow trees or caves. Barn owls often give a loud shriek when in flight. Females lay between four and seven eggs twice a year. It grows to a length of 32 cm (13 inches).

▼ **Elf owl** Only 15 cm (6 inches) long, this is one of the smallest owls. In Mexico and the USA the elf owl often nests in a deserted woodpecker hole in a saguaro cactus or a tree trunk. The male finds a suitable site and then sings to attract a female. Male and female elf owls sing a duet of yips, whines and barks. They come out at dusk to hunt insects such as moths, grasshoppers, beetles and scorpions. Occasionally they catch small snakes or lizards. When a male brings food to his mate they both rock their bodies.

▼ **Frogmouth** In Tasmania and parts of Australia this poor flier catches prey at night by jumping on to it from a tree. Its diet includes insects, snails, frogs and small mammals. A frogmouth spends the day resting. Its mottled plumage blends very well with the branches covered in lichen. At any hint of danger it stretches out its body, with head and bill pointing upward. Sometimes a frogmouth adopts an aggressive posture. It opens its beak wide in a huge gape, fluffs up its feathers and claps its wings. A flimsy platform of sticks makes a nest for this 45 cm (18 inch) long bird.

▶ **Sandgrouse** The male **chestnut-bellied sandgrouse** uses his chest feathers to absorb water like a sponge. Then he flies up to 40 km (25 miles) back to his chicks who suck the feathers to drink. This is vital for the chicks' survival in the hot and dry African deserts where they live. The main food of this species is plant seeds which they extract with their sharp bills. Huge flocks of chestnut-bellied sandgrouse gather morning and evening. They circle high above any water they find, land and fly off after drinking for a few seconds. A male sandgrouse is 32 cm (13 inches) long.

◀ **Kingfisher** With its bright plumage and long bill this 'king' among fishers lives near inland waterways and marshes in Europe, Asia and Africa. It perches on branches overhanging water or flies low over the surface. Plunging straight into the water it catches fish and insects. Male and female kingfishers look alike and are 15 cm (6 inches) long. Both partners dig a burrow in the riverbank. They start by hurling themselves at the bank again and again so that their sharp bills strike the same spot. Loose soil is scraped out with their feet. The tunnel slopes slightly upward and has a nesting chamber at the end. Four to eight eggs are laid each year.

▶ **Wood duck** or **Carolina duck** This North American species builds its nest in a tree away from water. The treehole is often as high as 12 m (40 feet) above ground. This is a strange location for a duck to choose for a nest. It is probably to stop predators eating its eggs. Wood ducks line the hole with plant material and feathers. As many as 12 eggs are laid in the nest by the female. The chicks hatch out in spring and soon after, they jump out of the nest. They follow their mother and glide down to the ground. Wood ducklings do not learn to fly until they are about six weeks old. They feed on nuts and other plant matter. Adults are 45 cm (18 inches) long.

◀ **Parrots** have big heads, short necks and bright plumage. Their upper beak is hooked and hinged with strong muscles. They feed on nuts, seeds, fruit and nectar. Parrots are related to the budgerigar, *kakapo* and *macaw.* **Grey parrots** are the best known African birds. They can imitate human speech and the calls of other birds. Flocks of grey parrots roost together in tall trees. They fly off at sunrise, squawking and whistling, to look for food. Females lay 2–4 eggs in a hole high up in a tree. An adult is 30 cm (1 foot) from beak to tip of tail. They live for up to 80 years.

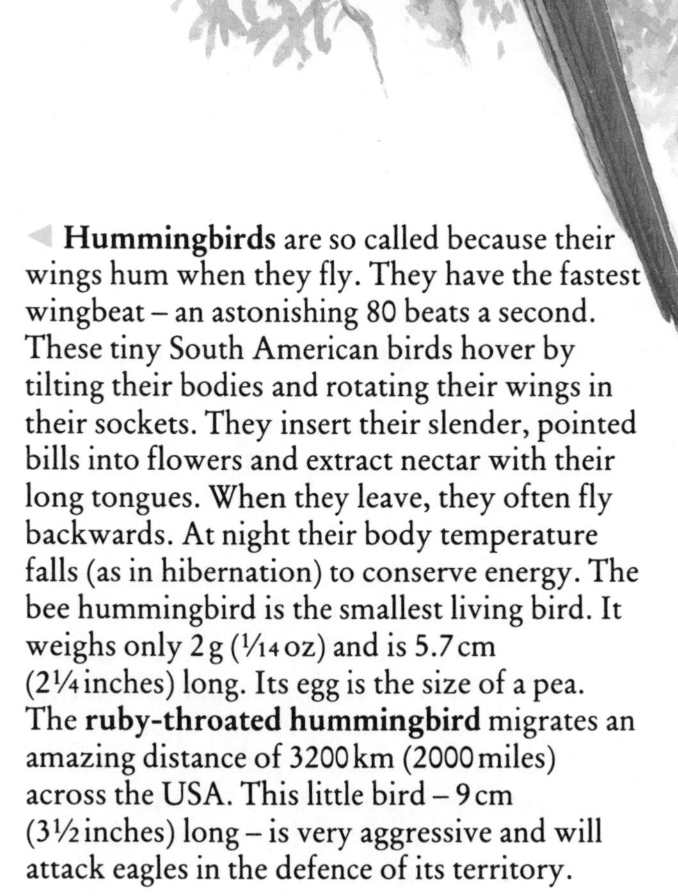

▶ **Macaws** have the same kind of feet as *parrots* with two toes at the front, two behind. This helps them to grip branches. Their hooked beak is so strong that they can split Brazil nuts with it. One of the largest members of the parrot family is the **scarlet macaw.** It grows to a length of 85 cm (34 inches), including its very long tail. The number of scarlet macaws in the wild is declining because of the widespread destruction of South American forests and because they are collected as cage-birds. Pairs of scarlet macaws fly together to their feeding grounds. Their 2–4 young stay in the tree-hole nest for about 13 weeks. When scarlet macaws are excited their facial skin 'blushes' pink.

bee hummingbird's egg

◀ **Hummingbirds** are so called because their wings hum when they fly. They have the fastest wingbeat – an astonishing 80 beats a second. These tiny South American birds hover by tilting their bodies and rotating their wings in their sockets. They insert their slender, pointed bills into flowers and extract nectar with their long tongues. When they leave, they often fly backwards. At night their body temperature falls (as in hibernation) to conserve energy. The bee hummingbird is the smallest living bird. It weighs only 2 g (1/14 oz) and is 5.7 cm (2¼ inches) long. Its egg is the size of a pea. The **ruby-throated hummingbird** migrates an amazing distance of 3200 km (2000 miles) across the USA. This little bird – 9 cm (3½ inches) long – is very aggressive and will attack eagles in the defence of its territory.

**Pheasants** are game-birds that scratch for food – seeds, shoots, berries and insects – in undergrowth. They roost in tree-tops. **Common pheasants** or **ring-necked pheasants** are found in forests in Asia, China and Japan. They have been introduced in Europe, North America and New Zealand. In spring the cocks (males) give courting displays of loud calls followed by bouts of wing rattling. They circle the hen (female) with a series of short hops. After mating, the hen scrapes a hollow nest and lays 8–15 eggs, one a day. Cock pheasants grow to a length of 90 cm (3 feet) of which the tail feathers measure 51 cm (20 inches).

**Partridges** live in small groups on lowland heaths throughout Europe. These game-birds like to run, although they fly up into trees to roost. When the fields and meadows are covered in snow in winter, partridges rely on food put out by humans. The **red-legged partridge** is distinguished by its red legs and white stripe over the eye. Cocks and hens look alike. The cock makes a well-hidden scrape lined with grass and leaves. The female usually lays 10–16 pale, yellow-brown eggs in the scrape. Sometimes the hen lays her eggs in other birds' nests. The young feed themselves on insects and green leaves. They are fully grown – 34 cm (13 inches) – at 16 weeks.

**Quails** seldom fly but are fast runners. Many **California quails** are hunted for sport and food. They are larger than European quails. The males are 30 cm (1 foot) long. They have strong bills with rough edges and they have a characteristic head plume. California quails move in flocks, mostly on foot. Their diet consists of leaves, seeds, berries and some insects, spiders and snails. The female lays 10–17 eggs in a leaf-lined nest. The mountain quail makes the shortest migration of any bird. It leaves its nest 3000 m (9500 feet) up in the Californian mountains in September and walks in groups of 10–30 down to the 1500 m (5000 feet) mark. In spring it goes back up the mountain to breed.

◀ **Guinea fowl** are African game-birds generally found in flocks in forests and bushy grasslands. They have bare necks and heads. Their dark feathers are speckled with white spots. The **helmeted guinea fowl** gets its name from the bony lump on its head. It drinks every evening before flying noisily up into a tree to roost. Guinea fowl have short wings. They run away when disturbed. Their varied diet includes insects, seeds, leaves and bulbs. Two females often share a scrape and each lays 10–20 eggs. Adults grow to a length of 60 cm (2 feet).

▶ **Turkeys** The **common turkey**, ancestor of the farm turkey, is found in the wild in North America and Mexico. Its legs are longer and its body is slimmer than farm breeds. Both sexes have bald heads and necks. When courting, the male turkey fans his tail, inflates the pouch over his bill and gobbles. His body is up to 1.2 m (4 feet) long. Females lay 8–15 eggs in shallow nests lined with leaves. The common turkey roosts in trees. It drinks twice daily and feeds on seeds, nuts, berries and insects. Strangely, thousands of farm turkeys freeze to death each year because they refuse to go into their warm sleeping quarters.

◀ **Chickens** have over 9000 feathers. The domestic hen is expected to lay an egg a day. These eggs are used in cooking. Chickens have three toes pointing forward and one pointing backward. The domestic chicken's ancestor is the **red junglefowl.** This bird from Asia is related to the *pheasant.* The female is smaller and duller than the male. She makes a scrape near a bush or bamboo clump, before laying 5–6 eggs. The male is 75 cm (2½ feet) long, including a 30 cm (1 foot) tail. Red junglefowl gather in groups of 50 or so to feed on grain, grass, berries and insects.

▶ **Swift** There are 67 species of this fast-flying bird. The alpine swift flies at 240 km/h (150 mph). Apart from nesting, swifts do everything while in flight: eat, drink, gather nesting material and mate. Alighting on vertical surfaces, such as buildings or cliffs, they grip with their four forward-pointing toes. Some species fly 24,000 km (15,000 miles) each year in migration. **Common swifts** breed in Europe and Asia, but winter in Africa. They utter a harsh cry as they fly in search of insects. Their cup-shaped nest of grass and feathers is glued together with saliva. Both parents feed their young but may leave for several days at a time.

◀ **Thrush** Over 300 species include *robins*, blackbirds and thrushes themselves. The **mistle thrush** – 27 cm (11 inches) long – is found in Europe, Russia and Asia. It migrates south every winter. Mistle thrushes spend a lot of time on the ground, running and hopping with the body held upright. When flying they fold back their wings at regular intervals, resulting in an up and down flight pattern. When the thrush is excited, a rattling chatter is heard. It eats fruit, earthworms and insects. A nest of moss, lichen and grass is built in a tree fork as high as 15 m (50 feet). Three to five eggs are laid in two broods. Young thrushes fly after about 15 days.

▶ **Robin** This popular bird is found in gardens and woodland in Europe, Asia, North America and North Africa. A **Eurasian robin** has a bright red breast. Female and male look alike and are 15 cm (6 inches) in length. They live alone during winter. Each robin defends its territory against intruders. Puffing out its red breast it sways from side to side. Robins feed on insect larvae, spiders, worms and fruit. The female builds a nest in a bush, tree or building. She lays as many as seven eggs two or three times a year. The young leave the nest after 15 days, even though not yet able to fly.

◄ **Swallows** fly swiftly and with great agility. They perch on telegraph wires and buildings, but rarely on trees. They catch insects in flight or pluck them from the surface of water. A long, forked tail and red forehead and throat are characteristic of swallows. **Barn swallows** are found in most parts of the world. In autumn in Europe huge flocks gather before migrating to Africa or India; North American swallows spend the winter in South America. When they arrive back in spring, both sexes construct a nest of mud and straw, often inside a building. The female lines the nest with feathers before laying 4 or 5 eggs. A fully grown barn swallow is 20 cm (8 inches) long.

► **Martins** and *swallows* look similar and have the same gliding flight. They catch their food – usually flies – while in the air, often over water. Martins have shorter wings, smaller forked tails and no red patch on forehead or throat. **House martins** gather beside puddles in spring to collect mud for their nests. They build their nests under eaves on the outside of buildings, under bridges or on cliff-faces. The nest is enclosed except for a narrow opening at the top. House martins have two or three broods a year. Their young fly when only 20 days old. Adults are 13 cm (5 inches) long and can fly at speeds of 80 km/h (50 mph).

◄ **Pigeons** live worldwide, except in Antartica. They have short necks, round bodies and a hump at the base of the bill. Two eggs are laid in a flimsy nest of sticks. The male incubates the eggs by day and the female takes over at night. The young are fed with 'pigeon's milk', full of protein and fat, from the parent's crop (throat). Homing pigeons can reach speeds of 100 km/h (60 mph). The ancestor of all domestic pigeons is the **rock dove**. This blue-grey bird is a ground feeder, eating berries, grain, seeds and sometimes snails. Rock doves are 30 cm (1 foot) long. They are found in Europe, India, Sri Lanka and north Africa.

▶ **Woodpeckers** are found worldwide. Their feet are adapted for clinging to tree trunks: their first and fourth toes point back; the second and third toes point forward. Their short, stiff tail acts as a prop while they listen for the sounds of insects moving within the bark. Then the woodpecker drills a hole with its pointed beak. It flicks its very long tongue in to collect its prey. In the USA deserted woodpeckers' nest are taken over by *elf owls*, screech owls and sparrowhawks. The male **great spotted woodpecker** drills a hole 30 cm (1 foot) deep in the trunk of a tree. The female lays 4–7 eggs in this unlined nest. Male and female make 40 trips a day to fetch nuts, seeds and berries for their young. This increases to 150 trips a day by the time the young are 10 days old. Adults grow to 25 cm (10 inches).

◀ **Cock-of-the-rock** This bird is found in South America. In the breeding season, groups of about 12 males assemble. When a drab brown female appears, the males flop down with a squawk and begin their displays. Crouching with their heads cocked to one side to show off their orange crests, the males then bounce up and down, making clicking sounds with their beaks. The **Andean cock-of-the-rock** lives in forests and nests on rocky cliffs. The female lays two eggs in a cone-shaped nest. These birds are 38 cm (15 inches) long and have sharp claws. They fly through the trees searching for fruit to eat. Females also catch frogs and lizards to feed to their young.

▶ **Toucans** All 37 species are found in South American forests. This fruit-eater has an amazing bill: long, thick, brightly coloured and slightly down-curved. It is used to push through dense foliage to reach fruit. A toucan picks figs one at a time, throws them into the air and catches them at the back of its throat. The bill is made of bony material which is light but strong. Toucans nest in holes in hollow trees. They have up to four young which are blind and naked when they hatch. The **toto toucan** is the largest toucan – 60 cm (2 feet) long. It will enter people's houses, steal food and tease pets.

◀ **Honeyeaters** are small birds found in Australia, on some Pacific islands and in southern Africa. These tree-dwellers are aggressive towards other species. Their slender, pointed beaks help them to eat fruit and insects. Honeyeaters have a long tongue frilled at the tip, like a brush. This helps them to pick up nectar from inside flowers. The base of the tongue is curled to form two long grooves along which the nectar is carried down the bird's throat. **Strong-billed honeyeaters** live in open forests in Tasmania. They are also known as bark-birds because they strip bark off eucalyptus trees in their search for insects. The bark is used in the building of a nest suspended from a drooping branch. One of the 167 species – the Kauai O-o – was thought to be extinct but was rediscovered in 1960 on a Hawaiian island.

▶ **Magpies**, members of the crow family, have a habit of collecting glittering objects. In the wild, magpies are shy and wary; but when tamed they make entertaining pets. They feed on insects, snails, slugs, spiders, grain, fruit, nuts and will steal young birds from other nests. Magpies are found on most continents and pairs choose a tree on a hillside or at the edge of a pond. They build a large domed nest of dry twigs and mud with an inner layer of hairs. The female lays 5–7 greeny/yellow eggs. A fully grown magpie has a 35 cm (14 inch) body and a 23 cm (9 inch) tail.

◀ **Weaver birds** are members of the sparrow family. Their name comes from their ability to tear strips from palm leaves and weave them into a hollow ball. The **sociable weaver** is found in dry scrub in Africa. Males and females look alike. Sociable weavers build an enormous nest in an acacia tree which houses as many as 300 pairs. Within the nest each pair of birds has its own chamber. They live and roost there all year round. Each female lays three or four dull white eggs. The same huge nest may last for 100 years, with constant repair and rebuilding. Sociable weavers leave the nest daily to feed on grass, seeds and insects. Adults are 15 cm (6 inches) long.

▶ **Nuthatches** are found throughout the world except in South America and New Zealand. These small birds have strong legs and sharp claws which help them climb up and down tree trunks. Nuthatches hunt face downwards for insects and spiders. They also eat seeds and nuts. They push a nut into a hole in the tree trunk and hammer with their beak to crack it open. The **European nuthatch** has a short tail. It hops along the ground and roosts in tree holes. The male feeds the female. They build a loose nest of bark and dead leaves in a tree hole, wall crevice or nestbox. Females lay 6–9 white eggs marked red and grey. The European nuthatch grows to about 15 cm (6 inches).

◀ **Larks** are widely distributed around the world. They have spread from Russia since the 14th century. You may sometimes see a pair of crested larks walking or running along a city road looking for seeds, buds and insects. They nest on the ground near rubbish dumps and highways. The nest is a scruffy mixture of stalks and roots. The female lays 3–5 eggs and the young leave the nest after 9–18 days. Male and female larks look alike, but the female is often smaller. **Skylarks** are 18 cm (7 inches) long, have a short crest and long, pointed wings. On each foot there is a long, straight hind claw. They enjoy dustbaths and singing early in the morning. Skylarks are found in Europe, Asia, Africa and have been introduced into Australasia and Canada.

▶ **Wagtails** have a long tail which wags up and down. These slender birds are found throughout the world. Like *nuthatches* and *larks*, wagtails have a long hind claw on each foot and like to walk along the ground. The **pied wagtail** of the British Isles has a cousin on the continent of Europe, the white wagtail. Both have similar black and white markings and develop a white throat in winter. They migrate to southern Africa and southern Asia for the winter. Pied and white wagtails nest on the river banks, in meadows and in buildings. Hundreds roost together in trees or reeds. Taking off from a fast, tail-bobbing run, they fly up and down eating insects in the air. Wagtails are 18 cm (7 inches) long.

◀ **Wood warblers** Most of the 125 species of wood warbler in America migrate at night. Black-and-white warblers fly across the USA at a steady 32 km (20 miles) a day. **Yellow warblers** live in damp thickets or swamps. The male has bright plumage, especially in the breeding season. The female is duller, with green feathers. Yellow warblers eat beetles, spiders and grasshoppers on the ground. Pairs build a nest in small trees or bushes, no more than 1.5 m (5 feet) above the ground. The nest is made of bark, plants, grass, feathers and cobwebs. Females lay four or five eggs and incubate them for 11 days. Adult yellow warblers are 13 cm (5 inches) long.

▶ **Finches** are seed-eating songbirds. They are widely distributed around the world. It is possible that thousands of years ago a flock of finches was blown by a freak storm away from the coast of South America. Blown across the Pacific Ocean, they found refuge on the Galapagos islands near Ecuador. The Galapagos finch uses its sharp beak to dig for insects in tree bark. Then it holds a cactus spine in its beak and dislodges the insect. The **American goldfinch** uses its beak for extracting tiny seeds from thistles. This small bird – 12 cm (5 inches) – builds a nest similar to that of the *yellow warbler*. Flocks of goldfinches form in the autumn and some migrate south for the winter.

◀ **Crossbills** are members of the *finch* family. They do not stay in one area but wander around searching for cones. They are called crossbills because the upper and lower parts of their beaks cross over. This enables them to prise and lever protein-rich seeds from their tough covering. The **common crossbill** of Europe and Asia is known as the **red crossbill** in North America. These acrobats hang and feed at any angle. Sometimes they tear off the cone and hold it in one foot. They are rarely seen on the ground. Three or four eggs are laid in a nest at the end of a branch. The young have symmetrical bills. Common and red crossbills are up to 15 cm (6 inches) long.

▲ **Red plumed bird of paradise** is one of the 42 known species of this spectacular bird found in Australia and New Guinea. The female is usually dull brown but the male shows off its bright plumage in a dazzling acrobatic display before the female. On a perch, it bends forward and raises its wings until they touch. Then lowering its head beneath its feet, it spreads out its side plumes. A male red plumed bird of paradise is about 35 cm (14 inches) in length.

▲ **King bird of paradise** Males of this species are 30 cm (1 foot) long, including the two wire-like tail feathers. The courtship display of a male involves raising its tail wires, hanging upside down and vibrating its wings. He also hops up and down on a branch, reversing his body at each hop. King birds of paradise feed on fruit. Females are well camouflaged in the dense forest. They are the only bird of paradise to make a nest in a treehole.

▶ **Iiwi** This is a member of the honeycreeper family. It is found near plants on the Pacific island of Hawaii. The iiwi feeds on nectar and insects, particularly caterpillars. It flies great distances to reach different areas as the various plants come into flower. An adult iiwi is 15 cm (6 inches) long – the same body length as the *king bird of paradise*. The iiwi is now endangered because people on Hawaii use its feathers to make cloaks for their tribal chief.

◄ **Tailorbird** The unusual nest of the **common tailorbird** is a cradle made from two leaves. It makes a series of holes with its beak and then pulls through strands of wool, spiders' web or cocoon silk to draw the edges together. Three or four eggs are laid on a bed of soft fibres. The adult tailorbird is 12 cm (4¾ inches) long. In the breeding season, the male's two central tail feathers grow longer than the others. Much time is spent hopping about in bushes, searching for insects and small spiders.

► **Hoopoe** The hoopoe is named after its 'hoop-hoop' call. Its crest lays flat and is only raised when the bird is excited. The hoopoe walks and runs swiftly, probing for large insects with its long bill. It flies slowly up into a tree to roost. Six or seven eggs are laid in a tree-hollow nest. Parents feed the nestlings until they are about 27 days old, at which time they leave the nest. Adult hoopoes are 28 cm (11 inches) long and are found in Europe and Asia.

◄ **Hornbills** The 44 species occur in Africa and tropical Asia. Most hornbills have a huge but lightweight bill, topped with a horny projection. These omnivores eat fruit, insects, lizards and other small animals. Hornbills are best known for their strange nesting habit. When a female **great Indian hornbill** is ready to lay her eggs, she hides in a treehole. The male seals the entrance with mud, leaving a small, slit-like opening. The eggs are safe from snakes and monkeys. During incubation and the first few months of the chicks' life, the male brings food. His long, curved bill is ideal for passing food through the slit. After the female breaks out, both parents bring food until the chicks are able to fly. A male great Indian hornbill is 1.5 m (5 feet) long from beak tip to tail – bigger than most children.

1 **Gannets** are found over oceans around the world. They soar above water, searching for fish and squid. Once prey is sighted, they plunge-dive, folding back their wings as they enter the water with a neat splash. Their streamlined bodies have no external nostrils. Large colonies of thousands of gannets breed on rocks. Each pair defends its own small territory. The nests are so closely packed that incubating birds are near enough to touch one another. The single chick has brownish feathers and develops adult plumage over a few years. A **northern gannet's** body is 90 cm (3 feet) long.

2 **Herring gulls** occur along coasts and inland waters of North America and Europe. They catch fish, scavenge on waste, steal other birds' eggs, prey on young birds and small mammals. Herring gulls also follow a plough, eating worms and other invertebrates. Their yellow bill has a red spot on it. Males and females look alike, but males are often longer – up to 60 cm (2 feet) long. In the breeding season, colonies are found on cliff slopes and beaches. The herring gull's nest is usually made of weeds and grass in a hollow in the ground or sometimes in a tree or building.

3 **Razorbills** have distinctive white bill markings. In winter the cheeks and forehead change to white. Parties of razorbills fly low over the sea or sit on rocks. They feed on fish, crustaceans and molluscs, using their wings for propulsion under water. On rocky shores along the Atlantic coasts of Europe and North America, pairs display by rattling their bills. The single chick leaves the nest after 15 days. Adult male razorbills are 42 cm (16 inches) long.

4 **Albatrosses** are found over southern oceans. Their spectacular gliding flight often lasts for hours without a wingbeat. The **wandering albatross** has the greatest wingspan of any living bird: 3.2 m (10½ feet). This seabird weighs about 8 kg (18 lb). Its diet consists of squid, fish and the refuse thrown overboard from ships. The albatross only comes to land to breed – every other year. As this large bird cannot flap its wings rapidly enough to take off, it usually tumbles off a cliff and uses warm air currents to gain uplift.

②

①

5 **Guillemots** are similar in many ways to *razorbills.* They spend most of the time swimming and diving offshore in the North Atlantic and North Pacific Oceans. They migrate south in winter. Guillemots fly just above the waves with a rapid whirring of their short wings. They call to one another with a growling 'arrr'. At the end of May, the female lays a single egg on top of a rock or cliff. Each egg is different in colour and pattern. This helps the adults to recognise their own egg in a crowded breeding colony. The **common guillemot** is 43 cm (17 inches) long when fully grown.

6 **Puffins** fly with very quick beats of their short wings. They dive for fish, sometimes returning to the surface with several hanging from the beak. **Common Atlantic puffins** live on rocky coasts around the North Atlantic Ocean. In winter the famous striped bill is smaller and duller. Puffins walk clumsily on land because their legs are set well back on the body. They nest in a burrow abandoned by another bird or a rabbit. If no burrow is available, the female digs a new one with her webbed feet. A single egg is laid each year. An adult common Atlantic puffin is 37 cm (15 inches) long.

1 **Ducks** are waterfowl, like *geese* and *swans*, with an oil gland situated near the base of the tail. A duck transfers oil to its feathers with its bill. This helps to waterproof the duck's body. Drakes (males) have brighter plumage than females. **Pintail ducks** are found in the northern hemisphere. They flock south to coasts, estuaries and inland waters for winter. The white neck and forked tail make pintail ducks easy to recognise in flight. They walk well on land and usually feed on vegetation at night. Pintails are generally silent. They often breed on islands in lakes and marshes. The scrape is lined with soft down. Seven to nine eggs are laid. Young chicks are tended by the female. The body of an adult pintail is 35 cm (14 inches) long.

2 **Geese** are water birds with short legs, webbed toes and broad bills. **Canada geese** breed in Canada and Alaska. Winter is spent in Europe, North Africa or southern USA. When they migrate, the same route is used by generation after generation. Females in particular return to their own birthplace to breed. On longer flights Canada geese fly in V formation. These birds live in a variety of habitats, from arctic tundra to rain forest. They also vary in size – from 56 cm–1 m (22–39 inches). A Canada goose feeds on grass and water plants. It calls with a loud, trumpeting honk. The nest is a shallow scrape, lined with down and plant matter. While the male stays nearby to defend his mate, she incubates her clutch of five or six eggs for about 30 days.

3 **Terns** are usually seen skimming over waves and sand dunes in flocks. The body rises and falls with each wingbeat. Terns seldom fly at a great height. After leaving their nesting-ground for the first time, sooty terns do not alight on land until they return at the age of three or four. The **common tern** is distinguished by the black tip to its red bill. This water bird breeds in North America, Europe and Asia on coasts, rivers and lakes. Common terns hover over prey, then dive rapidly into the water to seize small fish with their bills. Colonies of hundreds of thousands breed on isolated beaches, islands or cliffs. Pairs scrape a hollow in the ground, line it with vegetation and three eggs are laid. Common terns are 35 cm (14 inches) long.

◄ **Plover** These small wading birds have short legs, a round head and a sharp bill. The **ringed plover** is found on shores, inland waters or tundra in Iceland, Europe and Russia. It runs and stops short to dip quickly at insects, molluscs and worms. When nervous, a ringed plover bobs its head. The female makes a scrape in sand, shingle or turf and sometimes lines it with pebbles. Four eggs are laid in two or three broods each year. When a predator approaches the nest, the ringed plover flaps its wings as if injured and unable to fly. It gradually leads the predator away. Once clear of the nest, the plover flies up and escapes. Adult plovers are about 20 cm (8 inches) long.

► **Redshank** This is the noisiest of all shore birds. It makes calls of 'tuuu' and 'tu tu tu'. Redshanks push their long bills into mud, searching for worms and shrimps. They often wade in shallow water, bobbing their heads. These shy and restless birds are always on the move. After flying, redshanks keep their wings raised. They arrive on marshes and moorlands in Europe and Russia in April and May. They pair off after courtship displays. Females lay four eggs in a cup-shaped nest on the ground. Redshanks migrate south to estuaries and shores in northern Africa and southern Asia for winter. Males and females look alike and are 28 cm (11 inches) long.

◄ **Flamingoes** Huge numbers of these tall birds live and breed in colonies. All five species have very long legs and necks. When a flamingo flies, it moves in an up and down line as it beats its wings slowly. As it stands in water, it lowers its head and holds its bill beneath the surface. By moving its fleshy tongue, water is pushed through the bill and tiny food particles are sieved out. The **greater flamingo** only breeds every two or three years. Females scoop mud up in their bills to build cup-shaped nests on Caribbean or Galapagos islands. Usually only one egg is laid and both parents take turns to sit on it. The young chick is dependent on its parents for about 70 days, until the filtering mechanism in its bill is developed and it can fly. Males stand 1.5 m (5 feet) tall.

▲ **Jacana** The eight species are all long-legged water birds found worldwide. Their feet are adapted for their habit of walking on floating lily pads. Their toes and claws are so long that their weight is spread over a large area which stops them sinking. Jacanas eat plants, seeds and insects. They swim and dive well, but fly slowly. Males and females look alike, although females are slightly larger. **American jacanas** grow to a length of 25 cm (10 inches). Their nest is an untidy raft of plants on lakes or ponds in the USA and South America. Usually four buff-brown eggs are laid in the nest twice a year.

▲ **Avocet** This long-legged wader has an unusual bill – long, thin and up-curved. An avocet feeds on the insects and small aquatic animals that filter through its bill as it sweeps its head from side to side in shallow water. In deeper water the avocet dips its head below the surface and 'up ends' like a duck. An avocet often rests by standing on one leg. In flight, it retracts its neck and its legs project beyond its tail. Flocks of avocets stay together all year. They nest in simple scrapes. Males and females incubate four eggs and tend the young for at least six weeks. Adults are 40 cm (16 inches) long.

◀ **Frigate birds** are huge black birds with a 2.4 m (8 foot) wingspan and a deeply forked tail. They skim at high speed close to the surface of the sea and, with a nod of the head, pluck *squid* or *flying fish* from the water with their hooked bill. They will torment a returning booby bird so much that it drops its catch of fish. The frigate birds swoop on the dropped fishes before they hit the water. They spend most of the year in the air, seldom settling on water. In the breeding season, the male inflates his scarlet throat pouch. The **magnificent frigate bird** is also known as the man-o'-war bird. It is the fastest flying sea-bird, reaching 154 km/h (96 mph).

▲ **Whimbrel** This wader belongs to a group known as curlews. Males and females look alike. They are 40 cm (16 inches) long. Their long, down-curved bill is used to probe for small creatures, such as insects and worms. Whimbrels breed on moors in Canada, Asia and northern Europe. As the snows disperse, males begin their courting displays. The female lays four eggs in a nest usually in the open. Both the male and female incubate the eggs for four weeks. When the chicks are fully fledged, the adults migrate south and the young follow a few weeks later. They spend the winter in Africa, South America or Australia.

▲ **Oystercatchers** are noisy coastal birds found virtually worldwide. Their long, blunt, scarlet bill is used to prise open shellfish. They also eat insects and worms. If attacked, oystercatchers will run along the shore, swim or fly away. Huge flocks of oystercatchers migrate but return to the same area each year. In Norway a pair of oystercatchers returned to the same nesting place for 21 years. The **common oystercatcher** is 45 cm (18 inches) long. It has slightly webbed feet. A hole in the ground, lined with grass and moss, serves as a nest for 2–4 eggs. The chicks leave the nest the day after they hatch.

▶ **Swans** Whistling swans and **Bewick's swans** are very similar. They have short legs and webbed feet. Most of their time is spent on water, feeding on plant and animal food. Both Bewick's and whistling swans breed in the far north and migrate tremendous distances to winter in Europe, China, Japan and the USA. Males and females look alike, although females are slightly smaller than the 1.4 m (4½ foot) male. Pairs of swans stay together permanently. The female lays her clutch of 3–5 eggs in a nest of sedge and moss lined with soft feathers. This nest is usually near the water. The female incubates the eggs for up to six weeks. Young swans have mottled grey plumage.

1 **Pelicans** are found on lakes and coasts in Europe, Asia and Africa. As the pelican pushes its head under water, the lower bill expands into a large pouch and fills with water and fish. As the pelican lifts its head, the pouch contracts, forcing out the water but retaining the fish. Groups of pelicans gather in a semi-circle to feed together. They dip their bills to scoop up a whole shoal of fish. The **greater white pelican** has a yellow patch on its breast. Large numbers of them breed in colonies. One or two eggs are laid in a nest of sticks in a tree or on grass. Greater white pelicans grow to 1.75 m (5¾ feet).

2 **Herons** have slim bodies, large wings and long necks and legs. All 60 species have patches of powder-down feathers on their breast and rump. The powder is used in preening to remove slime from the bird's plumage. In Europe, Asia and Africa, the **grey heron** feeds mainly on fish and eels. As it wades through shallow water, it thrusts its head forward suddenly and grasps prey with its dagger-like beak. When resting, a grey heron often stands on one leg. A nest of sticks and reeds is built high in a tree. Three to five young herons hatch and chatter noisily, waiting for food. Adult grey herons are 90 cm (3 feet) long.

3 **Curlews** are very shy birds. In summer they breed on moorland, heath and grassland in Europe and Asia. They use the long bill to pick fruit and berries. Female curlews make large, lined scrapes in heather or reeds. Four glossy eggs are laid and both parents feed and tend the young. In winter curlews migrate south to mudflats. They use their down-curved bill to probe the mud for crustaceans and molluscs. They perch on rocks and posts. Their name comes from their call: 'cur-lew'. Adult males have a bill about 13 cm (5 inches) long. Their total length is 60 cm (2 feet).

4 **Storks** fly with head and neck stretched out and legs trailing down. **White storks** breed in eastern Europe and west Asia, often in forests near human habitation. Their nest takes eight days to construct. This large structure is built from sticks in a tree or on a building. As parents change over shifts on the nest they perform a greeting ceremony. They turn their heads round over their backs and clatter their bills. Storks feed mostly on frogs, reptiles and molluscs. They migrate south to Africa for winter. An adult white stork is 1.2 m (4 feet) tall. Its wingspan is nearly 2.1 m (7 feet).

5 **Mallards** are one of the most common *duck* species. They are found throughout the northern hemisphere, often feeding tail-up in water. Their diet includes grass, shoots and insects. Drakes (males) have bright plumage with a white collar and females have plain brownish plumage. Pair bonds are renewed each year with a prolonged display. Eight to ten eggs are laid in a nest on the ground or in a tree. Drakes fly off during the incubation period and undergo their annual moult. After hatching, the ducklings go on to water as soon as their feathers are dry. An adult mallard is about 60 cm (2 feet) long.

6 **Turnstones** breed on marshes near the Arctic coasts. Their plumage is bolder in summer (as shown here). Turnstones feed during the breeding season on plant material and insects, especially midges. The male chases the female during courtship. Four eggs are laid in a grass-lined hollow on stony shores. Both parents incubate the eggs. Turnstones fly south and spend the winter on rocky coasts. They turn stones and seaweed with their bills to find molluscs and crustaceans. As they forage they call out, 'kitititit'. Adults are up to 23 cm (9 inches) long. Turnstones are related to the *redshank*.

1
2
3
4
5
6

▲ **Ostrich** The largest living bird, 2.75 m (9 feet) tall and weighing about 127 kg (280 lb) is too big to fly but it is the fastest animal on two legs. It is adapted for running with powerful legs and two toes on each foot. It can run at 48 km/h (30 mph) for 20 minutes without tiring. In Africa an ostrich outruns its enemies, is watchful and never sleeps for more than 15 minutes at a time. The female produces the largest egg of any bird, equal in size to 40 chicken's eggs. Their chicks run as soon as they are born. Ostriches swallow stones to help them digest plants.

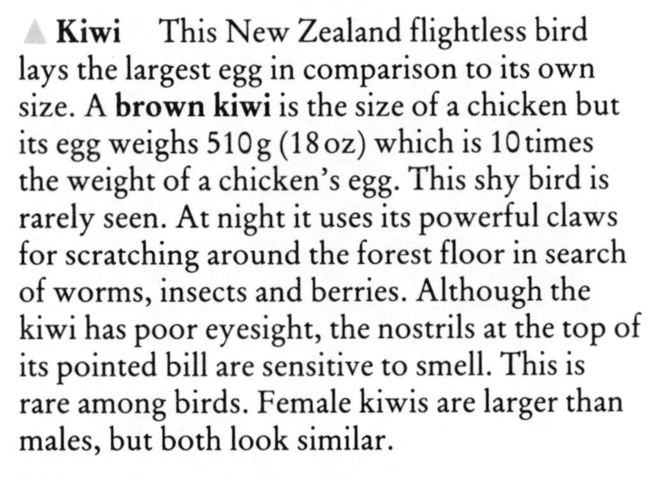

▲ **Kiwi** This New Zealand flightless bird lays the largest egg in comparison to its own size. A **brown kiwi** is the size of a chicken but its egg weighs 510 g (18 oz) which is 10 times the weight of a chicken's egg. This shy bird is rarely seen. At night it uses its powerful claws for scratching around the forest floor in search of worms, insects and berries. Although the kiwi has poor eyesight, the nostrils at the top of its pointed bill are sensitive to smell. This is rare among birds. Female kiwis are larger than males, but both look similar.

◀ **Penguins** are flightless birds whose wings are modified for propulsion underwater. Their webbed feet act as rudders. Short feathers form a dense, waterproof layer that provides insulation keeping them warm in their very cold habitat. Penguins stand upright and when alarmed slide on their bellies. **Emperor penguins**, the largest marine birds, are 1.2 m (4 feet) tall and weigh 16 kg (35 lb). Emperor penguins breed on pack ice in the Antarctic. Each male incubates an egg on its feet, protected by a flap of skin. Huddling together for warmth during the very cold, totally dark winter, the male penguins do not feed for 64 days.

▲ **Emu** The second largest living bird, the emu is 2 m (6½ feet) tall and weighs 54 kg (120 lb). The sole surviving species roams the plains of Australia. The emu runs nearly as fast as the *ostrich* and swims well. Large numbers have been killed as they were thought to be pests. The female tramples flat an area under cover before laying nine dark green eggs which have a pimply texture. The male emu incubates the eggs. After about 60 days, pale grey, striped chicks hatch. They leave the nest after two days and search for fruit, berries and insects.

▲ **Rheas** look very similar to *emus*. Like *emus*, although they have large wings, rheas cannot fly. Flocks of 20–30 **common rheas** of South America feed on plants, seeds and insects. Males and females look alike. At breeding time, each male displays to several females – running, roaring and jerking his neck and wings. Then he leads them to a shallow nest he has prepared. Each female lays a golden yellow egg. The male incubates as many eggs as he can cover. He charges and drives away any creature that comes near. Common rheas stand 1.5 m (5 feet) tall.

▶ **Cassowary** This large, flightless bird found in Australia and Indonesia is well adapted to forest life. Its quill feathers protect it from the undergrowth. As it searches for nuts, fruits and berries, the cassowary pushes forward the horny projection on its head to clear a way through. The cassowary often swims across streams. If threatened, it makes frightening booming calls that can be heard far away. Its legs are very powerful and it can inflict serious wounds with its sharp-toed feet. The female, larger than the male, is 1.5 m (5 feet) tall. She lays three to six eggs in a shallow nest lined with leaves.

◀ **Kakapo** Also known as the owl parrot, this species lives in New Zealand mountain forests. With land clearance by settlers and the introduction of predators such as stoats and rats, the kakapo is in danger of extinction. This unusual parrot cannot fly. During the day it shelters among rocks or in burrows. At dusk it emerges to feed on vegetation. A kakapo uses its beak and feet to climb trees. It flaps its wings to help balance and to glide down. Males are 60 cm (2 feet) long. They dig bowl-shaped areas to amplify their calls during courtship.

◀ **Takahe** This flightless bird was first discovered in New Zealand in 1849. Only 100 pairs were left alive in 1977 due to competition for plant-food from deer and because of their inability to fly away from predators. A takahe holds down a clump of snow-grass with one foot, cuts out stems with its beak and eats the tender bases. A nest of grass is made on the ground. Many eggs are eaten by predators or destroyed in bad weather. Usually only one chick from each clutch is reared. The takahe is 60 cm (2 feet) long, the same size as the *kakapo*.

▼ **Roadrunners** are among the fastest birds on the ground. They rarely fly. They run along roads, reaching 42 km/h (26 mph) when chased by a predator. In deserts and open country in Mexico and Arizona, USA, **greater roadrunners** feed on insects, lizards and snakes. They kill by making a sudden pounce. Pairs of roadrunners live in the same territory all year round. Two or three eggs are laid in a nest of sticks. Chicks are born naked and with eyes shut. If food is scarce, larger chicks take all there is and smaller ones starve to death.

# ANIMAL LINKS

So far we have covered the animal kingdom by looking at each animal separately within its own section, whether mammal, reptile, amphibian, fish, insect or bird. Now we are going to look at the different places where animals live. This will show the links between completely different species of animal as well as the links between an animal and its environment.

Each area of the world has its own collection of animal life. In many cases the animals of a particular area have had to adapt, often in ingenious ways, to survive. The hopping mouse, for example, touches the hot surface of the desert where it lives, only very briefly as it hops on narrow feet from one place to another. Other animals have had to adapt to a changing environment. Through the process of evolution over millions of years, fish that found their seas becoming overcrowded, developed ways of surviving on land. These became the first amphibians.

There are also links between the animals in a particular area. If you look at what they eat, you find a food chain connecting them. A caterpillar eats a leaf; a beetle eats the caterpillar; shrews feed on insects such as beetles; some snakes eat rodents such as the shrew; the snake in turn may become the food of a predatory bird like an eagle. In this way, animals are linked to one another to survive.

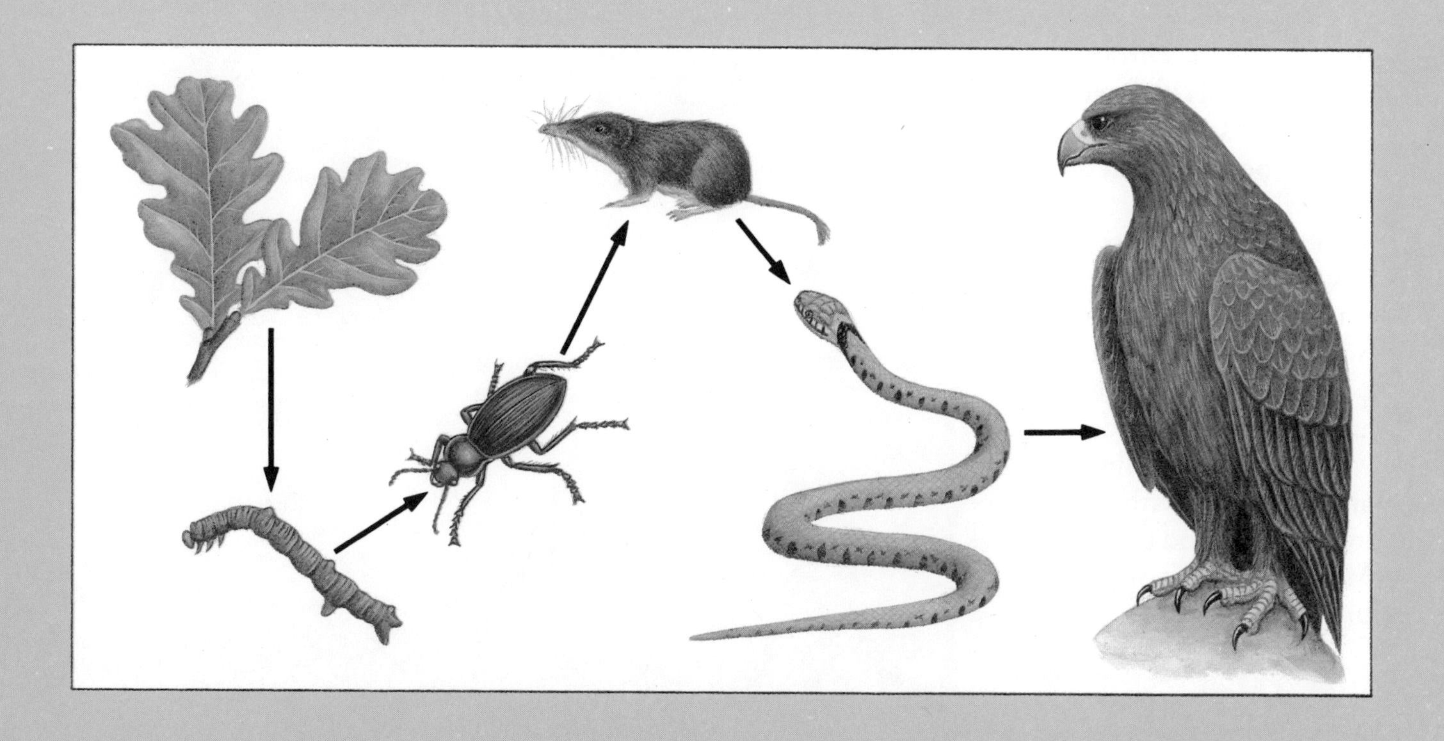

# In the Sea

Oceans and seas cover 70 per cent of the Earth's surface. The open sea is divided into layers, each with different types of animal life. There are many links between the creatures living here.

Sea birds fly above the surface. The wide span of their wings allows them to hover and soar. They plunge down to catch fish on or near the water surface.

Floating near the surface are jellyfish and millions of tiny plankton. These are the diet of many larger sea animals. Most animal life is found near the coast where the waters are shallow. Not all of the creatures shown here will be found in the same stretch of sea. This is because very few living things range over all oceans.

Some animals live only in warm seas, like the Atlantic eel, which is found in warm North Atlantic currents. Other animals like a certain amount of salt in the water, or need to be near light, or prefer the dark.

In the depths there are fewer creatures. They tend to eat each other because there is no plant life. At the very bottom, no light penetrates and the animals are dark in colour and often blind. They are adapted to withstand the intense cold and pressure. Many have light organs to lure prey.

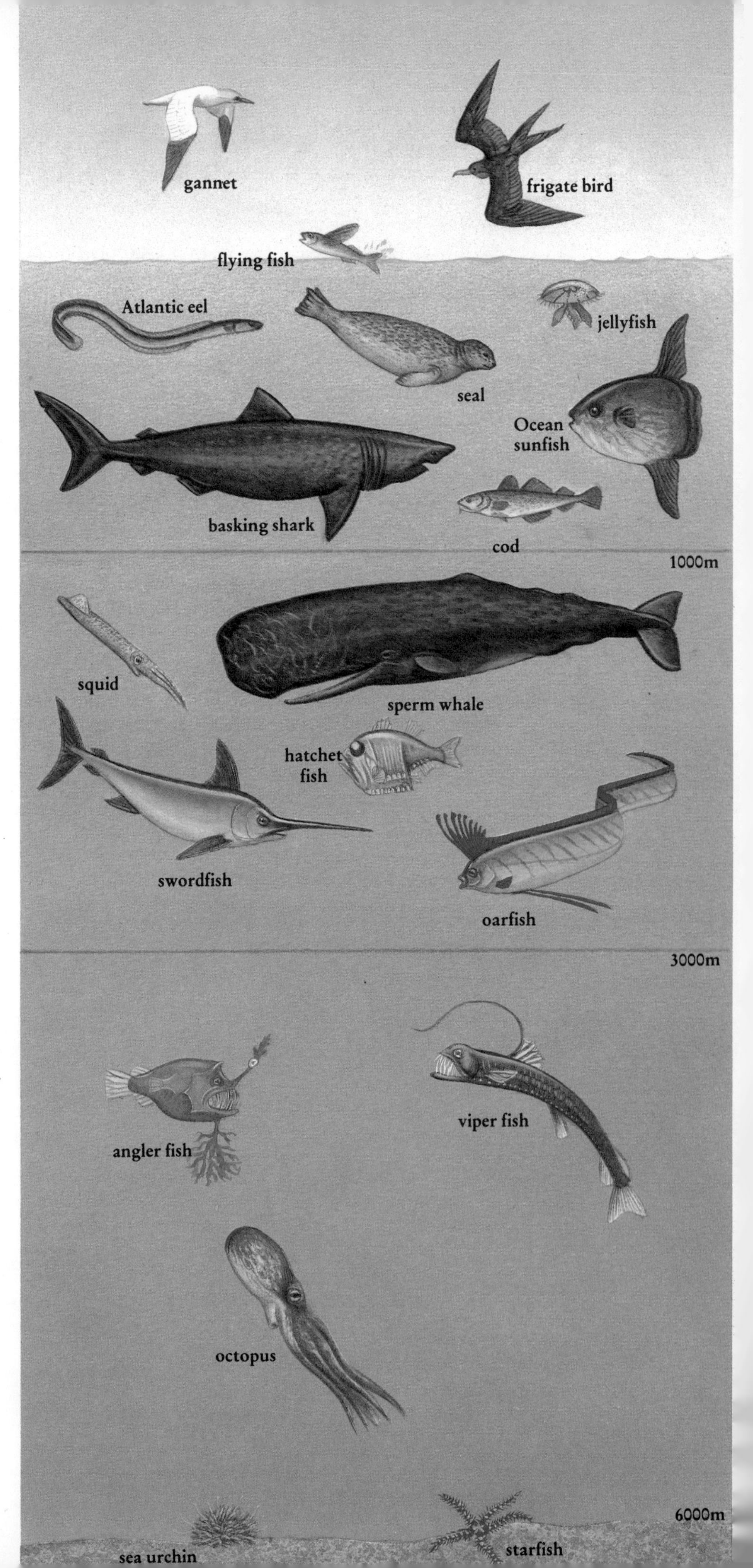

# On the Margins of the Land

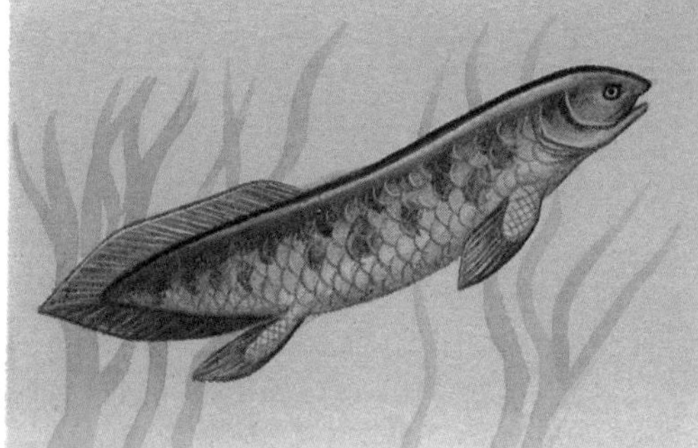

**Ceratodus** (extinct fish) could breathe air if water became too murky.

**Ichthyostega** (extinct amphibian) spent much of its time in water.

It was on the margins of the land (areas between the water and land) that the vertebrate land animals of today first appeared. Certain fish began to haul themselves out of the water 350 million years ago. Some came in search of water because their pools were drying up; others came in search of food because there were too many creatures in the sea. They had to overcome two main problems: how to move once out of the water and how to breathe oxygen from the air. Fossils of a primitive bony fish, the *coelacanth*, show that its fleshy fins enabled it to move on land. The *mudskipper* today moves on land with the aid of fleshy fins. The extinct **ceratodus** lived in fresh water about 200 million years ago. It could breathe air if necessary, like the modern *lungfish*.

The first group of vertebrate fish to adapt to life on land became the amphibians. Their structure is midway between fish and reptiles. They developed two pairs of limbs, nostrils and lungs. Unlike reptiles, amphibians have moist skin which is used in breathing. Most amphibians, like *frogs* and *newts* need to live in moist areas; all adult amphibians return to water to breed. Some of the early amphibians were large and frightening. **Ichthyostega** grew to 90 cm (3 feet) and had jaws spiked with sharp teeth.

## Sea margins

From the sea came the first back-boned creatures to live on land. They had to adapt to the land in the way they moved and the way they breathed.

## River margins

The river estuaries also attract animals that have adapted to the conditions – too salty for most freshwater animals, not salty enough for most marine creatures.

# Deep in the Forest

Tropical rain forests are found near the Equator where the rainfall is over 200 cm (78 inches) a year compared with 58 cm (23 inches) in London, England or 17 mm (½ inch) in the Sahara Desert in Africa. Rain forests are mainly in central South America, Madagascar, central and west Africa and south-east Asia.The temperature varies little from the hottest to the coldest months. These hot, steamy places are full of lush vegetation and numerous species of animal.

This picture of a tropical rain forest in South America shows some of the animals that live there. Rain forests can be divided into layers. Each layer is an independent area. In the topmost branches of the highest trees live birds like the rare **harpy eagle** ① and some **butterfly** species ②. The next layer is called the canopy. Birds such as **toucans** ③ eat the fruits and **hummingbirds** ④ feed on nectar in the flowers. Many mammals have adapted to life high in the forest. In the middle layer, **spider monkeys** ⑤ use a prehensile tail to hold on to the branches while the **tree sloth** ⑥ grips with long claws and **flying squirrels** ⑦ glide from tree to tree. Many birds, butterflies, **bats** ⑧ and some snakes like the **boa** ⑨ can also be found in the trees. On the ground such creatures as **armadillos** ⑩ **jaguars** ⑪ **ants** ⑫ and **spiders** ⑬ are found. In the rivers live reptiles like the **crocodile** ⑭ and fish such as the **piranha** ⑮. Most of the activity in a rain forest takes place at dawn, dusk and during the night.

# Across the Forests and Grasslands

Vegetation once covered nearly half of the Earth's land surface. Despite the influence of humans using land to grow crops and to build cities, there remain large areas of natural vegetation. For example, a huge band of coniferous forest runs across the northern land areas of North America, Europe and Asia. Further south are areas of temperate forest and grassland. In the coniferous forests animals have adapted to living in cold temperatures, low rainfall and poor soil conditions. In the grasslands, such as the Russian steppes, the South American pampas, the North American prairies and the African and Australian savannahs, the animals have to survive in an environment of wide open spaces and grasses. Both grassland and forest support herbivores that feed on the vegetation as well as carnivores that feed on the herbivores. These two pictures each show examples of a grazer, a burrower and a predator.

## Northern coniferous forest

**Red squirrels** feed on conifer seeds and other plant life. They move fast to avoid predators such as the **goshawk**. The vole is also preyed on by hawks and must burrow underground to escape. **Voles** can travel under the snow for hundreds of yards before they have to come out for food.

## African savannah

Predators like the **cheetah** are important because they keep the number of plant-eaters at a level that can be supported by the vegetation. Grazers like the **zebra** can easily be seen in the wide-open spaces but they are fast runners and stay in herds for protection. Burrowers, such as the **mole rat**, can hide among the plants when danger approaches.

# In the Dry Desert

Deserts are areas of very little rain. Some are hot all year round (eg Sahara Desert in Africa); others are hot in summer and cold in winter (eg Gobi Desert in Asia). Animals have adapted to the desert environment, for example the thick outer skeleton of many insects prevents them from losing much body fluid. Some desert animals feed on dry seeds and do not need to drink. Some, like the termites in the Mojave Desert in the USA, appear for only a few days in winter and then disappear underground.

**Camels** survive for long periods without water.

The smallest of the foxes, the **fennec fox** of North Africa, shelters in a burrow in the sand. It hunts small animals at night. Its large ears are alert to any sound.

In some African deserts, the male **sandgrouse** ferries water for up to 40 km (25 miles) to its chicks. Its feathers absorb water like a sponge. The chicks suck the feathers to drink.

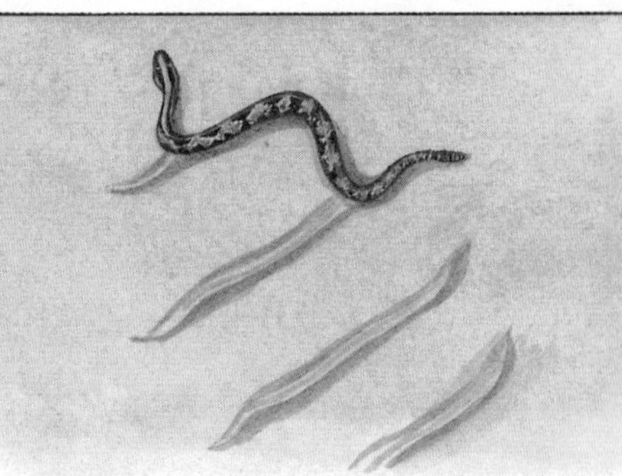

The **sidewinder** rattlesnake is found in the Nevada Desert, USA. It moves with a sideways motion, touching the hot sand with only two sections of its body at any one time.

**Spadefoot toads** lie buried in the Arizona Desert, USA, for 10 months each year. When rain falls they struggle out, lay eggs and feed as fast as they can for the long starvation ahead.

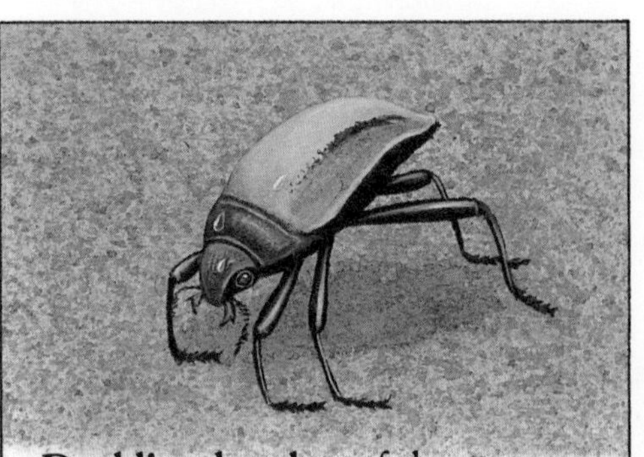

**Darkling beetles**, of the Namib Desert in Africa, collect moisture by standing head down in sand dunes as fog rolls in from the sea. The fog condenses into droplets on their bodies and trickles into their mouths.

The **hopping mouse** of central Australia has strong hind legs that enable it to hop away from predators. Only its narrow feet touch the hot sand. It emerges at night to feed on seeds and vegetation.

# The Frozen Wastes

Like the deserts, the frozen wastes are large areas of extreme temperatures – here extreme cold rather than extreme heat. No animals can live among the frozen peaks at the tops of mountains because there is little oxygen to breathe, but in the cold and desolate Arctic and Antarctic, some animals do exist. They have to face long hours of darkness and poor supply of food. They become part of a food chain depending on one another to survive.

Tundra is the wet, arctic grassland of North America, Greenland, parts of Scandinavia and the USSR. The frozen soil is called permafrost because it never thaws. Some animals of the tundra live there all year round, like the *lemmings* that tunnel under the snow. Others move north as summer arrives. Snowy owls and ravens prey on the lemmings. Dunlins and *turnstones* feed on insects. Arctic foxes feed on the chicks of these birds while herds of *reindeer* and elk graze on leaves and lichens. They migrate south again when the short summer is over.

**Arctic**
Harp seals feed on fish and squid in the sea but in their turn become prey for the polar bear.

**Antarctic**
Fish like cod is the main diet of the penguin which in its turn becomes the victim of the leopard seal.

# MAP OF T

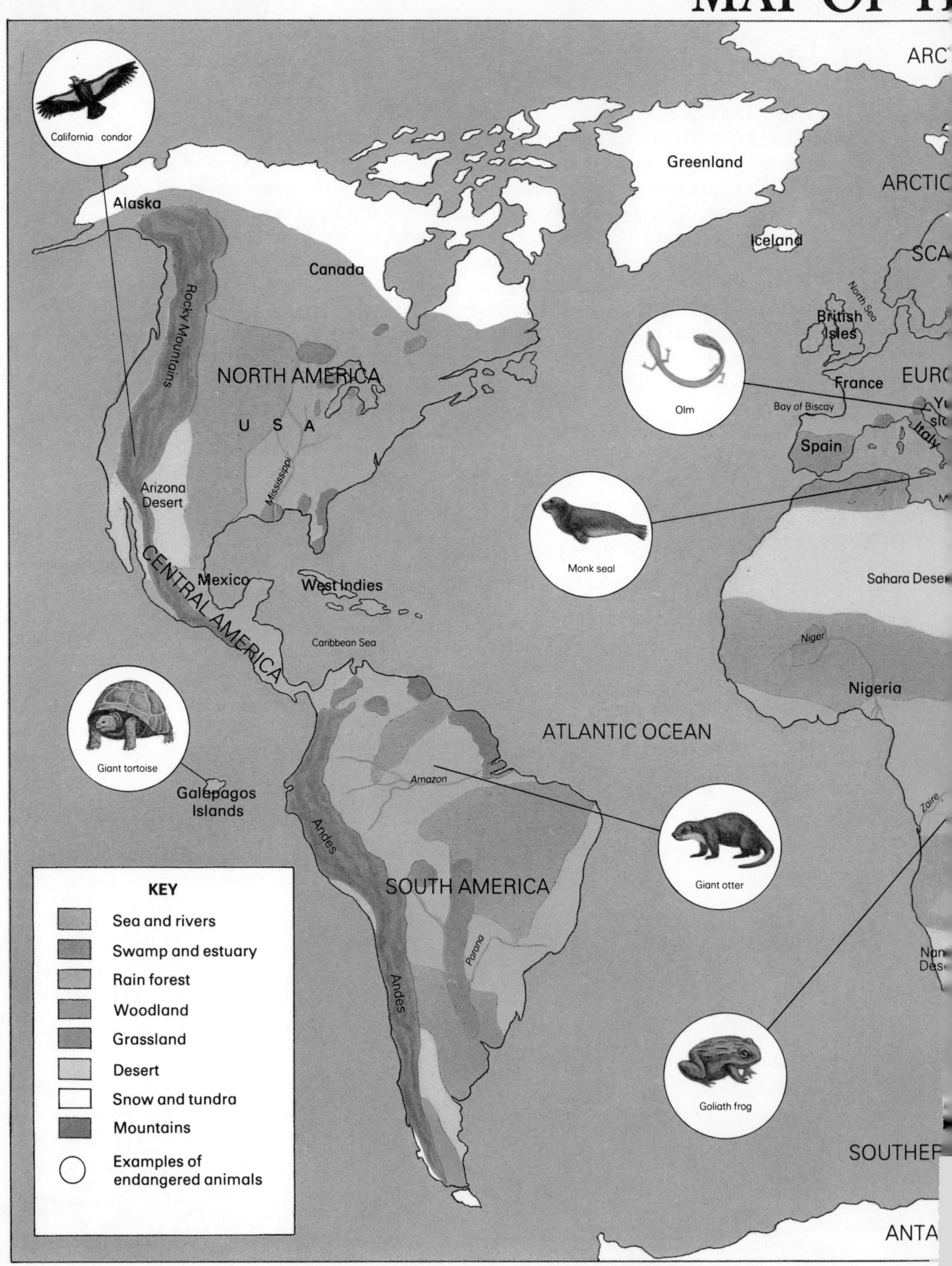

California condor
Alaska
Canada
Greenland
Iceland
Rocky Mountains
NORTH AMERICA
U S A
Mississippi
Arizona Desert
CENTRAL AMERICA
Mexico
West Indies
Caribbean Sea
North Sea
British Isles
France
Bay of Biscay
Spain
Italy
Olm
Monk seal
Sahara Dese
Niger
Nigeria
ATLANTIC OCEAN
Giant tortoise
Galapagos Islands
Amazon
Andes
Giant otter
SOUTH AMERICA
Parana
Zaire
Goliath frog
ARC
ARCTIC
SCA
EUR
SOUTHER
ANTA
KEY
Sea and rivers
Swamp and estuary
Rain forest
Woodland
Grassland
Desert
Snow and tundra
Mountains
Examples of endangered animals

# HE WORLD

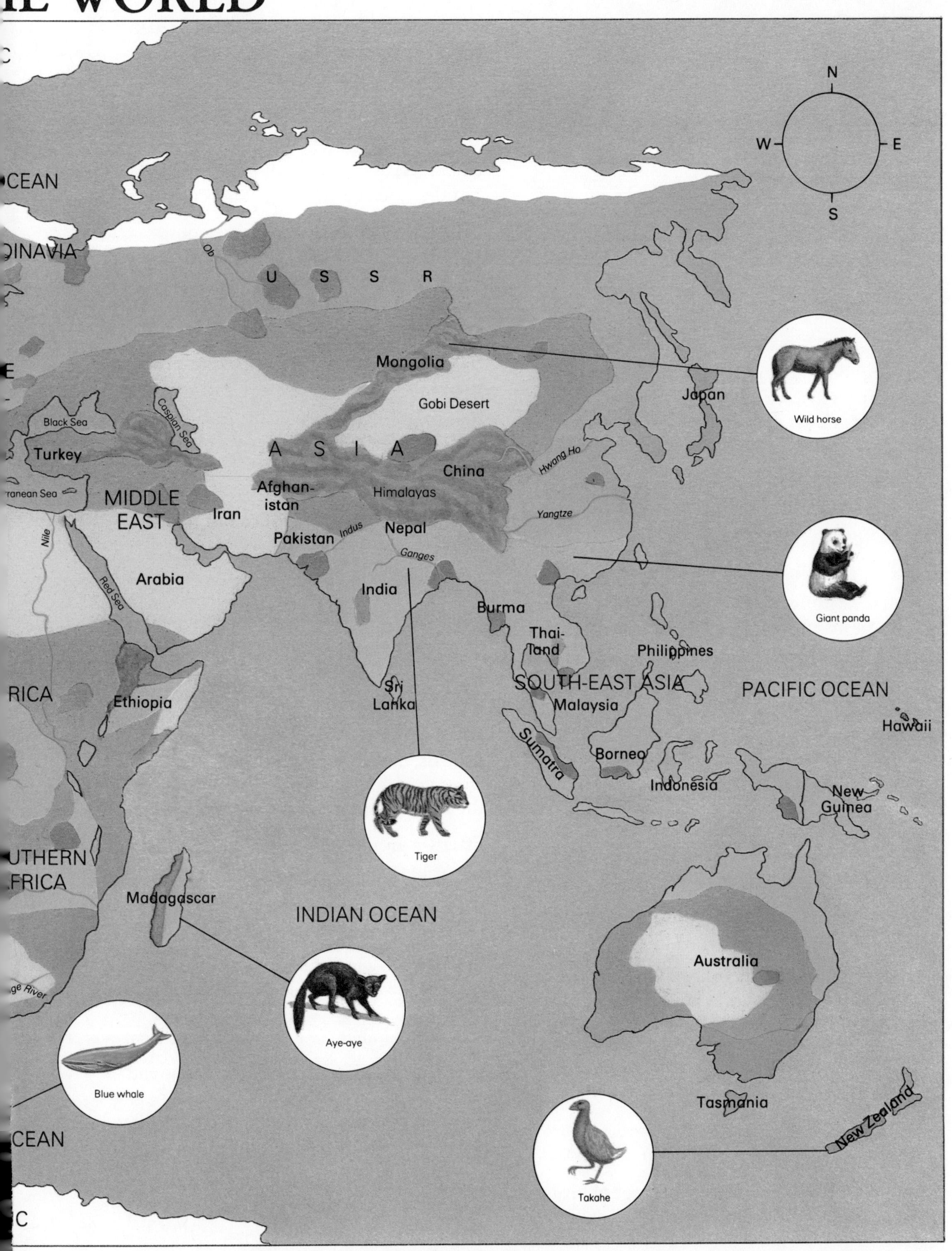

N
W
E
S
OCEAN
DINAVIA
U S S R
Ob
Mongolia
Gobi Desert
Japan
Wild horse
Black Sea
Caspian Sea
Turkey
A S I A
China
Hwang Ho
ranean Sea
MIDDLE EAST
Iran
Afghan-istan
Himalayas
Yangtze
Pakistan
Indus
Nepal
Ganges
Nile
Red Sea
Arabia
India
Giant panda
Burma
Thai-land
Philippines
SOUTH-EAST ASIA
PACIFIC OCEAN
Sri Lanka
Malaysia
RICA
Ethiopia
Hawaii
Sumatra
Borneo
Indonesia
New Guinea
Tiger
UTHERN AFRICA
Madagascar
INDIAN OCEAN
Australia
Aye-aye
Blue whale
Tasmania
New Zealand
Takahe
CEAN
C

# GLOSSARY

**abdomen** – in mammals, the belly; in insects, the third and rear division of the body; in arachnids, the body apart from the head.

**ancestor** – the creature of long ago from which others of today are descended.

**antennae** – feelers on an arthropod's head which are sensitive to touch, heat and smell.

**aquatic** – living in or near water.

**arachnids** – spiders, scorpions and mites.

**arthropods** – a group of invertebrates that consists of insects, arachnids, centipedes, millipedes and crustaceans. 80% of the animal kingdom are arthropods.

**barbel** – a piece of flesh hanging from the mouth of some fish, sometimes with a light on it.

**bovid** – a member of the ox family.

**brood** – a group of young animals (particularly birds or insects) hatched from several eggs at one time.

**budding** – how sponges and corals reproduce: the parent separates and one part grows into a new animal.

**camouflage** – a form of disguise that hides a creature in its habitat: colour or shape helps it blend in with its background.

**carnivore** – an animal that eats other living creatures.

**cephalopod** – a group of molluscs that includes squid and octopus.

**cocoon** – a case of silky threads made by an insect in its larval stage to hold the pupa.

**crustaceans** – arthropods with hard shells such as lobsters, crabs, barnacles and lice.

**domesticated** – tamed to work or to be a pet; not a wild animal.

**dorsal** – the fin on the back of some fish.

**down** – the fine, soft under-plumage of birds to keep them warm; it is also used to line nests.

**echinoderm** – a group of sea animals with spiny skin and symmetrical bodies (e.g. sea urchin and starfish).

**elongate** – something that is longer than it is wide (e.g. the skull of a horse).

**endangered** – more are dying than are being born and the species is in danger of becoming extinct.

**extinct** – when a species has died out.

**fry** – young fish once they are hatched.

**gill** – an organ in most fish and some amphibians for taking in oxygen from water.

**habitat** – the home or place where an animal is found.

**herbivore** – an animal that feeds on plants.

**hibernation** – a period of deep sleep to avoid the extreme cold of winter.

**imago** – the adult form of an insect after metamorphosis.

**incubation** – the act of sitting on eggs.

**insulation** – something that stops loss of body heat (e.g. a layer of fat).

**invertebrate** – an animal that has no backbone (e.g. insects, spiders and worms).

**larva** – the stage of development in many animals, particularly insects, when they hatch from the egg; the larva is usually different from the adult in shape and size. (Plural – **larvae**.)

**litter** – a group of young animals born at one time.

**mandible** – the paired parts behind the mouth used for biting and chewing. The jaw or jawbone in mammals; the beak in birds.

**marsupial** – a mammal with a pouch on the belly of the female. The young crawl into the pouch where they suckle milk and complete their development (e.g. kangaroo, wombat).

**metamorphosis** – the change of shape and form of a creature from larva to adult (e.g. tadpoles into frogs or toads, caterpillars into moths or butterflies).

**migration** – movement from one habitat to another, especially birds and fish which come and go with the seasons.

**mollusc** – a soft-bodied animal often with a hard shell. Most are slow-moving (e.g. snail, mussel, octopus).

**monotremes** – egg-laying mammals (platypus and spiny anteater). After hatching, the young feed on milk from the mother.

**moult** – the casting off of one skin as another grows in its place.

**muzzle** – the nose and mouth of some animals (e.g. dog and horse).

**nocturnal** – active at night.

**northern hemisphere** – one half of the world from the Equator to the North Pole.

**nymph** – the larva of certain insects that is like the adult but smaller.

**omnivore** – an animal that feeds on both plants and animals.

**opposable** – a finger or toe that can move in a different direction from the other digits (like the human thumb), helping an animal to grasp things.

**ovoviviparous** – producing eggs that are hatched within the body of the parent and then born living.

**parasite** – an animal which lives in or upon another living creature and draws food from it (e.g. mosquito).

**pectoral** – the fin on the chest or underside of some fish.

**placentals** – mammals whose young feed and develop inside the mother's body and are born at an advanced stage of development.

**plankton** – drifting or swimming animals and plants which live in water and are moved by currents. Many are too small to be seen without a microscope.

**powder-down** – very fine down in a bird's plumage used for preening.

**predator** – a carnivore that preys upon other animals, birds, their eggs or their young.

**preening** – the tidying and trimming of feathers with the beak, necessary for a bird to fly.

**prehensile** – able to grasp hold of things. Often used when referring to a monkey's tail.

**proboscis** – the long, flexible snout of an animal (e.g. elephant's trunk); the mouthparts of insects and other invertebrates.

**pupa** – the third stage of metamorphosis such as the caterpillar in the cocoon before it becomes a butterfly or moth. (Plural – **pupae; pupate** – to become a pupa.)

**radula** – the 'toothed tongue' of some molluscs in the form of a horny strip with ridges or 'teeth' on its surface which rasp rather than cut food.

**roost** – the place chosen by a bird to rest or sleep.

**ruminant** – a mammal with a four-part stomach that chews the cud. Food goes through two parts of the stomach, returns to the mouth and is well chewed again before going to the other two parts of the stomach.

**scavenger** – an animal or bird that feeds on the flesh of creatures killed by another animal (e.g. hyena, vulture).

**secrete** – to produce a body fluid that serves some function (e.g. spiders secrete silk).

**sonar** – sound waves that bounce off solid objects (e.g. bats send out squeaks that rebound from trees and other objects back to their sensitive ears, helping to guide their flight).

**southern hemisphere** – one half of the world from the Equator to the South Pole.

**spawn** – the eggs laid by fish and some amphibians.

**spinneret** – the organ in spiders and silkworms that produces silk.

**stridulation** – a harsh, shrill noise produced by certain insects (e.g. cricket) rubbing one part of the body (file) over another part (scraper).

**tapetum** – part of the eye in certain animals (e.g. cat) which glows at night when a light shines on it.

**tundra** – the vast, flat, treeless regions in the north of the northern hemisphere with arctic climate and vegetation. The ground is frozen and covered in snow and ice for most of the year.

**vertebrate** – an animal that has a backbone (e.g. mammals, birds, reptiles, amphibians and fish).

# INDEX

# THE ANIMA

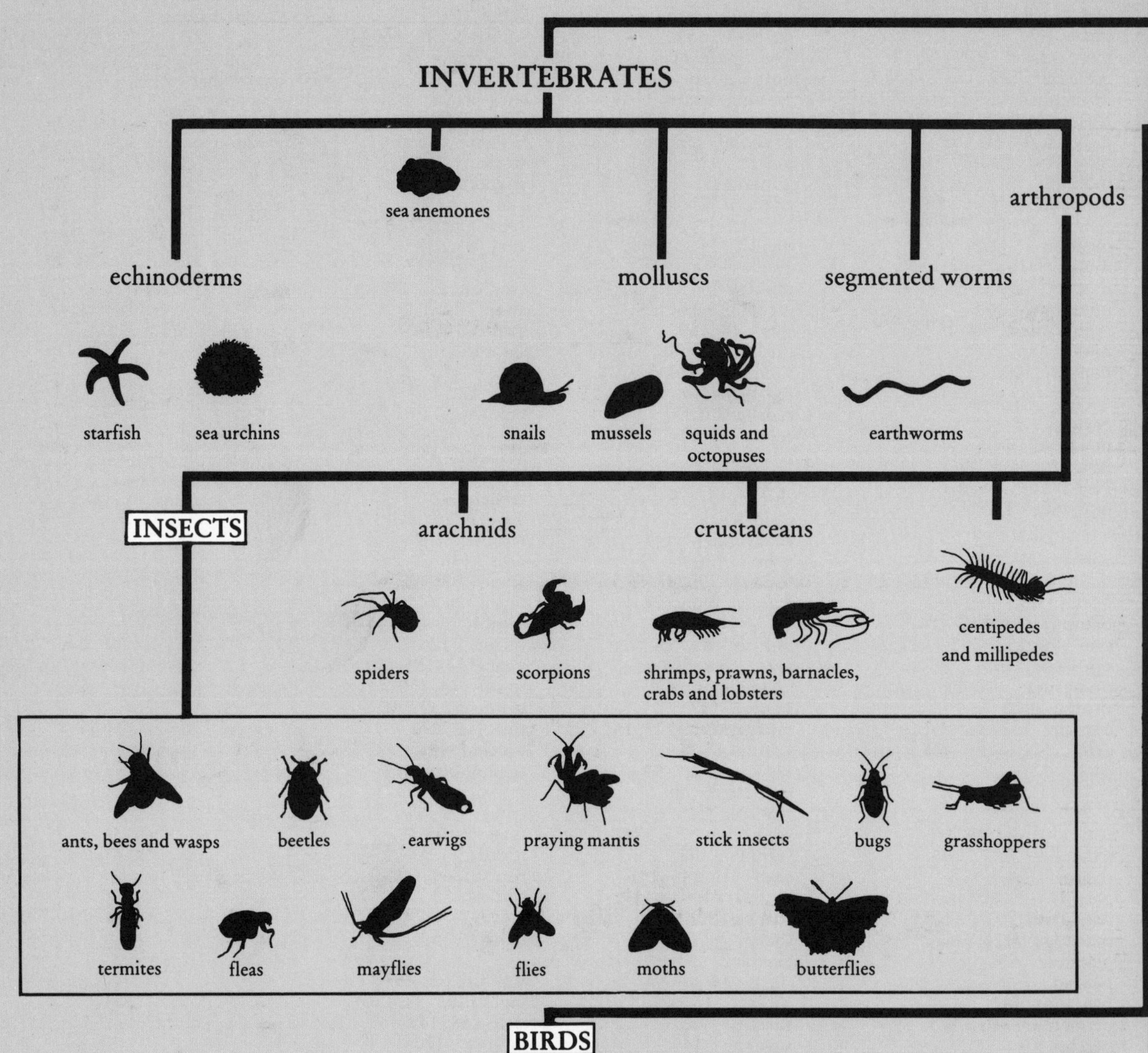

INVERTEBRATES
sea anemones
arthropods
echinoderms
molluscs
segmented worms
starfish
sea urchins
snails
mussels
squids and octopuses
earthworms
INSECTS
arachnids
crustaceans
spiders
scorpions
shrimps, prawns, barnacles, crabs and lobsters
centipedes and millipedes
ants, bees and wasps
beetles
earwigs
praying mantis
stick insects
bugs
grasshoppers
termites
fleas
mayflies
flies
moths
butterflies

BIRDS
eagles and hawks
vultures
owls
frogmouths
game-birds
kingfishers and hornbills
parrots
swifts and hummingbirds
pigeons
woodpeckers and toucans
albatrosses
waders and gulls
pelicans
herons, storks and flamingoes
ducks
ostrich
kiwis
penguins
takahe